METAL EXTRACTION BY
BACTERIAL OXIDATION OF MINERALS

ELLIS HORWOOD SERIES IN INORGANIC CHEMISTRY

Series Editor: Dr JOHN BURGESS, University of Leicester
Consulting Editor: Ellis Horwood, MBE

Inorganic chemistry is a flourishing discipline in its own right and also plays a key role in many areas of organometallic, physical, biological, and industrial chemistry. This series is developed to reflect these various aspects of the subject from all levels of undergraduate teaching into the upper bracket of research.

Alcock, N.W.	**BONDING AND STRUCTURE: Structural Principles in Inorganic and Organic Chemistry**
Almond, M.J.	**SHORT-LIVED MOLECULES**
Barrett, J.	**UNDERSTANDING INORGANIC CHEMISTRY: The Underlying Physical Principles**
Beer, P.	**HOSTS, GUESTS AND INCLUSION CHEMISTRY**
Burger, K.	**BIOCOORDINATION CHEMISTRY: Coordination Equilibria in Biologically Active Systems**
Burgess, J.	**IONS IN SOLUTIONS: Basic Principles of Chemical Interactions**
Burgess, J.	**METAL IONS IN SOLUTION**
Burgess, J.	**INORGANIC SOLUTION CHEMISTRY**
Cardin, D.J., Lappert, M.F. & Raston, C.L.	**CHEMISTRY OF ORGANO-ZIRCONIUM AND -HAFNIUM COMPOUNDS**
Caulcutt, R.	**DATA ANALYSIS IN THE CHEMICAL INDUSTRY: Volume 1, Basic Techniques**
Constable, E.C.	**METALS AND LIGAND REACTIVITY**
Crichton, R.R.	**INORGANIC BIOCHEMISTRY OF IRON METABOLISM**
Filov, V.A.	**HARMFUL CHEMICAL SUBSTANCES: Volume 1: Elements in Groups I–IV of the Periodic Table and their Inorganic Compounds**
Harrison, P.G.	**TIN OXIDE HANDBOOK**
Hartley, F.R., Burgess, C. & Alcock, R.M.	**SOLUTION EQUILIBRIA**
Hay, R.W.	**BIO-INORGANIC CHEMISTRY**
Hay, R.W.	**REACTION MECHANISMS OF METAL COMPLEXES**
Housecroft, C.E.	**BORANES AND METALLOBORANES: Structure, Bonding and Reactivity**
Kendrick, M.J., May, M.T. & Robinson, K.D.	**METALS IN BIOLOGICAL SYSTEMS**
Lappert, M.F., Sanger, A.R., Srivastava, R.C. & Power, P.P.	**METAL AND METALLOID AMIDES**
Lappin, G.	**REDOX MECHANISMS IN INORGANIC CHEMISTRY**
Maddock, A.	**MÖSSBAUER SPECTROSCOPY**
Massey, A.G.	**MAIN GROUP CHEMISTRY**
Massey, A.G.	**TRANSITION METAL CHEMISTRY**
McGowan, J. & Mellors, A.	**MOLECULAR VOLUMES IN CHEMISTRY AND BIOLOGY: Applications Including Partitioning and Toxicity**
Parish, R.V.	**NMR, NQR, EPR, AND MÖSSBAUER SPECTROSCOPY IN INORGANIC CHEMISTRY**
Raithby, R.R.	**TRANSITION METAL CLUSTER CARBONYLS: Structure, Synthesis, Reactivity**
Roche, L.P.	**THE CHEMICAL ELEMENTS: Chemistry, Physical Properties and Uses in Science and Industry**
Romanowski, W.	**HIGHLY DISPERSED METALS**
Snaith, R. & Edwards, P.	**LITHIUM AND ITS COMPOUNDS: Structures and Applications**
Tsitsishvili, G.V., Andronikashvili, G. Kirov, G.N., and Filizova, L.D.	**NATURAL ZEOLITES**
Williams, P.A.	**OXIDE ZONE GEOCHEMISTRY**

METAL EXTRACTION BY BACTERIAL OXIDATION OF MINERALS

JACK BARRETT
Formerly King's College London.
Gold Biohydrometallurgy Chemistry Consultancy,
Kingston-upon-Thames, Surrey

M. N. HUGHES
Department of Chemistry, King's College London

G. I. KARAVAIKO
Institute of Microbiology, Russian Federation
Academy of Sciences, Moscow

P. A. SPENCER
Spencer Hydrometallurgical Services, Perth, Australia

ELLIS HORWOOD
NEW YORK   LONDON   TORONTO   SYDNEY   TOKYO   SINGAPORE

First published in 1993 by
ELLIS HORWOOD LIMITED
Market Cross House, Cooper Street,
Chichester, West Sussex, PO19 1EB, England

A division of
Simon & Schuster International Group
A Paramount Communications Company

© Ellis Horwood Limited, 1993

All rights reserved. No part of this publication may be reproduced, stored in a retrieval system, or transmitted, in any form, or by any means, electronic, mechanical, photocopying, recording or otherwise, without the prior permission, in writing, of the publisher

Printed and bound in Great Britain
by Hartnolls, Bodmin

British Library Cataloguing in Publication Data

A catalogue record for this book is available from the British Library

ISBN 0-13-577735-6

Library of Congress Cataloging-in-Publication Data

Available from the publisher

Table of contents

Preface xi

1 Introduction
   1.1 The extraction of metals 1
   1.2 Refractory gold and roasting 1
   1.3 Other metals whose extraction involves roasting 2
   1.4 Alternatives to roasting 2
   1.5 The bacterially catalysed oxidation of arsenopyrite
       compared to roasting and pressure oxidation 3
   1.6 Introduction to bacterial oxidation 3
   1.7 Introduction to the bacteria 4
   1.8 Outline of the book 6
   1.9 References 7

2 Potentially treatable minerals
   2.1 Introduction 8
      2.1.1 Classification of bacterial oxidation reactions . . . 8
      2.1.2 Methods of carrying out bacterial oxidation reactions . . 9
   2.2 World production of some metals 12
      2.2.1 Gold 13
      2.2.2 Silver 18
      2.2.3 Cobalt 19
      2.2.4 Uranium 20
      2.2.5 Nickel 21
      2.2.6 Molybdenum 22
      2.2.7 Tin 23
      2.2.8 Copper 24
      2.2.9 Antimony 26
      2.2.10 Zinc 27
      2.2.11 Lead 27
   2.3 Conclusions 28
   2.4 Geochemical and chemical classifications 29
      2.4.1 Geochemical classification 29
      2.4.2 Chemical classifications 31

Table of contents

    2.5 References 36

3 The catalytic bacteria
    3.1 Introduction 38
    3.2 General microbiology 39
        3.2.1 The cell wall 41
        3.2.2 The cytoplasmic membrane 41
        3.2.3 The cytoplasm 41
        3.2.4 Capsules and slimes 42
        3.2.5 Systematization of bacterial species 42
        3.2.6 Metabolism of chemolithotrophs 42
    3.3 Bacterial Characteristics 43
        3.3.1 Genus *Thiobacillus* 44
        3.3.2 Genus *Leptospirillum* 48
        3.3.3 Moderately thermophilic bacteria 50
        3.3.4 Thermoacidophilic archaebacteria 53
        3.3.5 Guanine and cytosine content of bacterial DNA 55
    3.4 Catalytic activity of bacteria 56
        3.4.1 Interaction of bacteria with surfaces 57
        3.4.2 Factors influencing reaction rate 60
        3.4.3 Bacterial growth characteristics 62
        3.4.4 Conclusions 63
    3.5 Toxic effects of metal ions 63
    3.6 Mixed cultures 65
        3.6.1 The range of microorganisms in ores 66
        3.6.2 Mechanism of interaction between microorganisms . . . 66
    3.7 Isolation of bacterial cultures 68
        3.7.1 Isolation procedure 68
        3.7.2 Subculturing procedure 70
        3.7.3 Isolation of pure cultures 70
        3.7.4 Maintenance and storage of cultures 71
    3.8 References 71

4 The chemistry of bacterial oxidation reactions
    4.1 Introduction 73
    4.2 Aspects of pH measurement 74
    4.3 Speciation 76
        4.3.1 Sulfur(VI) 76
        4.3.2 Arsenic(III) 77
        4.3.3 Arsenic(V) 78
        4.3.4 Iron(II) 80
        4.3.5 Iron(III) 81
        4.3.6 Other solubilized metals 91
    4.4 Chemical equations 92
    4.5 Electrode potentials 92

4.5.1	The oxidizing power of oxygen	93
4.5.2	The oxidizing power of iron(III)	93
4.6	Thermodynamics of mineral oxidation	95
4.6.1	Oxidation of pyrite	95
4.6.2	Oxidation of chalcopyrite	96
4.6.3	Oxidation of arsenopyrite	97
4.6.4	Oxidation of arsenic(III)	98
4.6.5	Conclusions	100
4.7	Chemical oxidations	100
4.7.1	Chemical dissolution of chalcopyrite	100
4.7.2	Chemical dissolution of pyrite and arsenopyrite	101
4.8	References	101

5 The general mechanism of bacterial oxidation

5.1	Introduction	103
5.2	Previous mechanistic conclusions	103
5.3	The mechanism of bacterial oxidation reactions	105
5.3.1	Introduction to mechanistic principles	105
5.3.2	The mechanism of bacterial oxidation	106
5.3.3	The deduction of a general mechanism for the bacterial oxidation of pyrite/arsenopyrite concentrates	110
5.3.4	Mechanistic conclusions	118
5.4	Consequential predictions	122
5.4.1	pH	123
5.4.2	Iron(III) concentration	123
5.4.3	Arsenic(V) concentration	124
5.4.4	Arsenic(III) concentration	124
5.4.5	Iron(II) concentration	125
5.4.6	Electrode potential	125
5.4.7	Pulp density (solids density)	126
5.4.8	Bacterial population	126
5.5	References	126

6 Application of bacterial oxidation technology

6.1	Introduction	128
6.2	Fundamental operating conditions	128
6.2.1	Quantity of sulfide material	129
6.2.2	Aeration	131
6.2.3	Acidity	132
6.2.4	Temperature	132
6.2.5	Nutrients	133
6.2.6	Culture growth	133
6.3	Bacterial oxidation methods	134
6.3.1	Agitated reactor vessels	134
6.3.2	Heaps	141

Table of contents

    6.3.3 Dumps 148
    6.3.4 Vats 149
    6.3.5 *In situ* treatment of ore bodies 150
  6.4 Examples of processing methods 151
    6.4.1 Tank leaching 151
    6.4.2 Heap leaching 152
    6.4.3 Vat leaching 152
    6.4.4 Dump leaching 152
    6.4.5 *In situ* leaching 152
  6.5 References 153

7 Product and effluent treatment
  7.1 Products of bacterial oxidations 154
  7.2 Recovery of metal values 154
    7.2.1 Recovery of liberated gold 154
    7.2.2 Recovery of solubilized metal values 155
    7.2.3 Recovery of metal values from solid residues 156
  7.3 Clean-up of aqueous effluents 156
  7.4 Mineralogical and thermodynamic data 157
    7.4.1 Compounds containing iron(III) and/or sulfur(VI) . . . 157
    7.4.2 Compounds containing iron(III) with arsenic(V) and/or sulfur(VI) . 160
    7.4.3 Possible compounds representing ultimate
          thermodynamic stability 163
  7.5 Theoretical modelling of precipitate stabilities 163
  7.6 Experimental investigations of precipitated solids 165
  7.7 Conclusions 166
  7.8 References 168

8 Economic factors
  8.1 Introduction 169
  8.2 Capital costs 169
    8.2.1 Major equipment 170
    8.2.2 Materials of construction 170
    8.2.3 Ancillary services 170
  8.3 Operating costs 171
  8.4 Bacterial oxidation process costs 173
  8.5 References 177

9 Analytical methods
  9.1 Introduction 178
  9.2 Total concentrations of elements 179
    9.2.1 Atomic absorption spectroscopy:
          flame and electrothermal methods 179
    9.2.2 Atomic emission spectroscopy 180
    9.2.3 Inductively coupled plasma mass spectrometry 181

Table of contents

9.3 Measurements of specific chemical species 181
    9.3.1 pH measurement 181
    9.3.2 Polarographic methods 183
    9.3.3 Ultra-violet and visible techniques, including colorimetric methods 183
    9.3.4 Ion chromatography 183
    9.3.5 Gold analysis 184
9.4 Measurements of biomass 184
    9.4.1 General comments: measurements of biomass on and off the mineral 184
    9.4.2 Analysis of protein 185
    9.4.3 Differentiation between free cells and mineral-bound cells and between growing and non-growing bacteria 185
9.5 References 186

Index 187

Those were the early weeks I spent working at the Precipitation pits, and received seven shillings a week, the wages of a man.

Marvellous are the works of humans to take red water from a mountain and turn it into copper.

See them digging the new shaft on Parys - red-core Parys that is riddled with copper - working as did the Romans. Down, down goes the shaft into the bowels of the mountain: fill the shaft with water and leave it for nine months: then pump it out into Precipitation.

In come the ships carrying scrap iron for the stew.

Aye, this is the stew of copper - old bedsteads, old engines, wheels and nails - scrap of all description from the iron industry of the North comes surging into Amlwch on the ships. The carters unload it and carry it to the pits: and in it goes, iron scrap into the sulphate stew, the blood of Parys. And we, the stirrers, worked the brew with wooden poles - wooden ones, since metal ones would melt in the acid - even copper nails in our boots. Stir, stir, stir - nine months it took to melt the iron into a liquid. And the sludge that dropped to the bottom of the pits was copper. Break this up cold and cart it like biscuits down to the maw of the Smelter for refining.

This was the process; a job for men, not boys.

> This quotation is from *Land of my Fathers* by Alexander Cordell (Hodder & Stoughton Ltd, 1983), a fictionalized account of copper mining in Anglesey, North Wales, in the early part of the nineteenth century.

Preface

There is great interest at present in the application of microbiological methods to the extraction of metals. With regard to the primary extraction of metals from their ores and concentrates, bacterial oxidation methods are the most important. This book provides a critical discussion of the chemistry, mineralogy and environmental issues central to this subject, and includes a description of the appropriate microbiology. It is intended as a text for undergraduate and postgraduate students of mining engineering, mineral engineering, metallurgy, biotechnology, chemistry and microbiology. The study of the bacterially catalysed oxidation of minerals represents a major contribution to the interdisciplinary subject of biohydrometallurgy. We hope that this book will serve to introduce the subject to courses in the above disciplines and to engender further research activity. Additionally, we hope that it will encourage the mineral industry to adopt bacterial oxidation technology. At present, commercial application of these methods occurs on a relatively small scale and is restricted to the liberation of gold from refractory minerals and the leaching of copper and uranium. However, the potential for further application to a greater range of metals, as described in the text, is considerable. We believe that the economic advantages of these processes, coupled with the current emphasis on environmental protection, means that they will have a major role to play in the future development of industrial extractive metallurgy.

In a book of this length it has been necessary to be selective with the quoted references. It has not been possible to cover all the varied published contributions to the subject. Other information has been omitted because of its proprietary nature.

Although the authors have joint responsibility for the whole text, they wish to indicate the specialist inputs from Gregori Karavaiko, who wrote Chapter 3, and Peter Spencer who wrote Chapters 6 and 8. Duplication of the discussion of some aspects of the subject has been included deliberately to avoid excessive cross-referencing and to facilitate an uninterrupted flow of the text.

Preface

Acknowledgements

Two of us (J.B. and M.N.H.) wish to acknowledge the support of Bactech (Australia) Pty. Ltd., and Paragon Resources N.L., in funding the bacterial oxidation research group at King's College London, and the research collaboration with our co-author, Peter Spencer, together with Julia Budden of Bactech and Mike Rhodes of Paragon Resources. We are especially grateful for the early and continuing support from Joe Rotondella. We gratefully acknowledge our research collaboration with Robert Poole, who discovered the M4 mixed thermophilic culture which forms the basis of one of Bactech's commercial processess, and Des O'Reardon, Ali Nobar and Keith Ewart, who carried out much of the research work at King's College London.

One of us (J.B.) wishes to thank Bob Robins of Hydromet Technologies Ltd., for useful discussions and some unpublished data which were of assistance in the preparation of Chapters 4 and 7, and Ellis Horwood, Mike Shardlow and James Gillison for their support and advice in the preparation of the manuscript as camera ready copy.

Kingston upon Thames, U.K.	Jack Barrett
King's College London, U.K.	M.N. Hughes
Russian Federation Academy of Sciences, Moscow	G.I. Karavaiko
Perth, W. Australia	P.A. Spencer
October 1992	

1

Introduction

1.1 THE EXTRACTION OF METALS

The continuing supply of metals to support modern technology is of crucial significance for industry, the economy, the environment and society. It is now accepted that much greater concern must be exercised over environmental issues in the mining of ores and in the extraction of metals. There is also a need to develop more efficient processes that will allow the use of lower-grade ores and new processes to deal with ores that cannot be handled by conventional technology. Such advances are necessary to ensure the continued availability of a range of metals into the twenty-first century.

The central theme of this book is the application of new bacterial oxidation methods for the extraction of metals. Such methods offer the opportunity to develop economic, energy-efficient, pollution-free processes. This chapter presents an overview of these methods and some indication of the ores which may be treated by bacterial methods. These ores are usually those which are sulfidic in character (or which contain pyrite impurities) and are conventionally roasted as a pre-treatment stage before the metal is extracted. Although historically the extraction of copper was the first bacterially catalysed oxidation process (as indicated by the preliminary epigraph), it was the attraction of using it for the liberation of refractory gold that produced the research impetus that has led to its more general applications.

1.2 REFRACTORY GOLD AND ROASTING

Gold occurs naturally as the element. If the metal particles are small (in the region of -50 μm in diameter) they may have been rendered refractory by being encapsulated in minerals such as arsenopyrite, pyrite, pyrrhotite and chalcopyrite. Such encapsulation prevents cyanidation, the normal method of recovery of the metals, from being effective. The encapsulating minerals form an impervious physical barrier between the metal and the solution of sodium cyanide which, together with dissolved atmospheric oxygen as the oxidant, would normally dissolve both gold and silver and allow their efficient recovery. The conventional treatment of refractory ores and concentrates is to roast them in order to liberate the metals, the oxidant being atmospheric oxygen. Roasting results in the sulfur content of the mineral being oxidized to gaseous sulfur dioxide. Any arsenic content is oxidized to arsenic(III) oxide (arsenious oxide), As_2O_3, and the iron content is oxidized to iron(III) oxide (ferric oxide), Fe_2O_3. In the gaseous state arsenic(III) oxide consists of As_4O_6 molecules. The containment of the arsenic(III) oxide for essential environmental protection contributes greatly towards the costs of the process. The prevention of major atmospheric pollution by sulfur dioxide with its eventual conversion into 'acid rain' is also a major cost disadvantage to the process. The roasting method can be inefficient due to the re-encapsulation of the gold and silver by silicates.

1.3 OTHER METALS WHOSE EXTRACTION INVOLVES ROASTING

A considerable number of metals are found as sulfides and are normally extracted by roasting the ore or a concentrate. The production of these metals (which include copper, zinc, nickel, cobalt, antimony, tin, lead, bismuth, and molybdenum) has the associated problems of removal of sulfur dioxide and arsenic(III) oxide from the gaseous effluent.

1.4 ALTERNATIVES TO ROASTING

Other chemical methods of oxidizing minerals include the use of (i) gaseous oxygen at high pressures and temperatures (pressure oxidation), (ii) chlorine, (iii) nitric acid or (iv) sulfuric acid, but all suffer from plant complexity and non-competitive economics.

An example of a chemical method which has been proven by full-scale plant operation is pressure oxidation (Carvalho *et al*, 1988) which makes use of high pressures (typically 20 atmospheres) of oxygen (derived from a supply of liquid oxygen) which, at elevated temperatures (200°C), facilitate the appropriate mineral oxidations. In this case the solution products are sulfuric acid, arsenic(V) acid, H_3AsO_4 and iron(III) ion. The process does not lead to the production of atmospheric pollutants and shares the effluent problems of the alternative bacterial oxidation method. The severe conditions of pressure oxidation demand a high standard of plant design and materials, early commissioned plants having suffered failures in pumps, valves, impellers, pipes and gaskets, in addition to failure of the autoclave brick linings (Matthews, 1990).

Bacterial oxidation of minerals, the subject of this book, consists of the use of suitable bacteria to **catalyse** the reactions. In all cases the oxidizing agent is atmospheric oxygen and the reaction equations are identical with those describing pressure oxidation. Iron(III) acts as the oxidant in some reactions, but its reduction product, iron(II), is re-oxidized by oxygen.

1.5 THE BACTERIALLY CATALYSED OXIDATION OF ARSENOPYRITE COMPARED TO ROASTING AND PRESSURE OXIDATION

Taking the oxidation of arsenopyrite, FeAsS, as an example, the chemical equations representing its oxidation, by roasting or pressure oxidation and bacterial oxidation respectively, may be written as follows:

$$2FeAsS + 5O_2 \rightarrow Fe_2O_3 + As_2O_3 + 2SO_2 \tag{1.1}$$

$$2FeAsS + 7O_2 + 4H^+ + 2H_2O \rightarrow 2Fe^{3+} + 2H_3AsO_4 + 2HSO_4^- \tag{1.2}$$

Reaction (1.1) takes place at 600-800°C and is fairly rapid with typical residence times of eight hours. Reaction (1.2) takes place at ~200°C and ~20 atmospheres pressure (residence times between two to four hours) in the case of pressure oxidation and, owing to the catalytic effect of the bacteria, at 30-55°C and atmospheric pressure (residence times of four to six days) in the case of bacterial oxidation. Catalysts normally operate at very low concentrations and bacteria are no exception to this rule. The observed bacterial populations of operating systems are of the order of 10^{10} cells per millilitre (10^{13} cells per litre). Regarding the bacterial cell as a giant molecular assemblage, a molar concentration would consist of 6×10^{23} cells per litre (based on Avogadro's number) so that the operating population is the equivalent of a solution which is about 1.7×10^{-11} M in cells. This very low concentration is typical of that expected for catalytic substances. The details of the above chemical reactions are discussed fully in Chapter 4.

1.6 INTRODUCTION TO BACTERIAL OXIDATION

The use of bacterial oxidation for the liberation of gold (and any silver content) from refractory ores and concentrates forms the basis of a rapidly developing technology. The bacterial oxidation of sulfidic minerals occurs naturally and is a major cause of the acidity of mine waters. The closure of the Wheal Jane tin mine in Cornwall, UK, in 1991 produced a serious outflow of bacterially solubilized toxic metals into the River Fal. This resulted from acid mine drainage accumulating, and eventually escaping into the environment, when the mine was no longer being kept dry by pumping.

Bacterial oxidation is responsible for the solubilization of copper which is noticeable in sulfidic copper deposits, a process known (but not understood) by the Romans, and now used for the commercial production of copper (e.g. at the Bisbee and Bingham Canyon mines in the USA). Not until 1947 was it realized that the process involved bacteria when the bacterium known as *Thiobacillus ferrooxidans* was isolated (Colmer and Hinkle, 1947) from acidic mine water.

A general review of bacterial leaching has been published by Hutchins *et al.* (1986). A book by Hughes and Poole (1989) contains a chapter entitled 'Metals, micro-organisms and biotechnology' which consists of a general treatment of the subject. A major book by Rossi (1990) deals with all aspects of biohydrometallurgy and contains a large number of references. A book edited by Karavaiko et al. (1988) is also entirely devoted to biohydrometallurgy. A book edited by Ehrlich and Brierley, C.L., (1990) contains chapters by Norris, Tuovinen, Lawrence and McNulty, and Thompson which are relevant to the subject. The proceedings of two international symposia edited by Karavaiko and Groudev (1985) and Karavaiko et al. (1990) are devoted to the 'Modern aspects of microbiological hydrometallurgy' and the 'Dump and underground bacterial leaching of metals from ores' respectively. The proceedings of the biennial Biohydrometallurgy series of international symposia which appear every odd year are a useful source of information and discussion of all aspects of the subject. Bacterial oxidation technology features in the annually produced proceedings of the Randol Gold Forums.

In addition to its use for liberating gold from refractory ores and concentrates bacterial oxidation has two other general applications. One is the solubilization of metals from their sulfide ores and/or concentrates, the second being the solubilization of metals from oxide ores by making use of the products of bacterial oxidation of iron/sulfide minerals, i.e. aqueous iron(III) and sulfuric acid. The first of these applications allows the solubilization of cobalt from cobaltite, nickel from pentlandite, molybdenum from molybdenite, tin from stannite, copper from chalcopyrite, chalcocite and covellite, antimony from stibnite, and zinc from sphalerite.

The second application allows, for example, the solubilization of uranium from uraninite. The almost ubiquitous presence of pyrite in mineral deposits makes the separation of the two effects difficult. It is quite likely that the product of the bacterial oxidation of pyrite, iron(III), exercises an important intermediary role in the oxidation of what might be termed the 'non-ferrous' sulfides. This point is amplified in the discussion of the mechanism of the process in Chapter 5.

1.7 INTRODUCTION TO THE BACTERIA

It is useful, at this stage, to introduce some of the terminology of microbiology which applies to the catalytic bacteria. The bacteria which can catalyse mineral oxidation reactions are classed as **chemolithoautotrophs** (background microbiology given by Schlegel (1986)). A chemolithotroph is a bacterium which has the capacity to derive energy by oxidizing inorganic materials such as minerals. An autotroph is a bacterium which uses carbon dioxide as its sole source of carbon. The bacteria described in this section live a seemingly harsh life of chemolithoautotrophy! Some of the bacteria are strictly chemolithoautotrophic (obligate chemolithoautotrophs, i.e. they are obliged to be chemolithoautotrophs), while others, although they can metabolize organic compounds, have the additional facility to be autotrophic (facultative autotrophs) depending upon their particular environment. Heterotrophic bacteria make use of organic compounds as their carbon sources.

The bacteria are classified generally in terms of the temperature ranges where they can operate efficiently. There are four very broad classes, in such terms, as given in Table 1. Cryophiles are sometimes referred to as psychrophiles.

Table 1.1. Classification of chemolithotrophic bacteria
in terms of their optimum temperature ranges

Bacterial class	Optimum temp. range/°C
cryophiles	< 20
mesophiles	20-40
moderate thermophiles	40-55
extreme thermophiles	> 55

There are no cryophiles of interest to the current subject. There are at least three mesophiles which have suitable properties for sulfidic and arsenosulfidic mineral oxidation. The most commonly studied and used mesophiles are *Thiobacillus ferrooxidans, Thiobacillus thiooxidans* and *Leptospirillum ferrooxidans* which are obligate chemolithoautotrophs. The moderate thermophiles are not as well characterized as the mesophiles, only the genus *Sulfobacillus thermosulfidooxidans* (Golovacheva and Karavaiko, 1978) having been named, others being identified by codes such as TH-1, TH-2 and TH-3 (Brierley, J.A., and Le Roux, 1977, Brierley, J.A., et al. 1978, and Le Roux and Marshall, 1977) or M4 (Nobar et al. 1987). The *Sulfobacilli* and other moderate thermophiles are facultative chemolithotrophs. The extreme thermophiles belong to the genera *Sulfolobus* and *Acidanus* which are facultative chemolithotrophs. *Acidanus brierleyi* (Brierley, C.L., and Brierley, J.A., 1973, and Brierley, C.L., 1974) has the capacity to degrade sulfide minerals.

All the chemolithotrophs are regarded as being obligate aerobes in that oxygen, O_2, is essential as the ultimate electron acceptor or oxidizing agent. There is some evidence that some chemolithoautotrophs may, under appropriate conditions, exhibit facultative anaerobic properties in which they gain energy by using iron(III) as the oxidant. Under laboratory and plant conditions, in addition to oxygen and an appropriate sulfidic mineral, they require a nutrient solution (Silverman and Lundgren, 1959) containing ammonium, magnesium, calcium, potassium, chloride, sulfate, nitrate and phosphate ions. Traces of elements such as zinc and manganese are usually present in the mineral target materials in sufficient quantity and are not usually added to reacting systems. Under natural conditions the bacteria must have a source of the above nutrients in addition to a supply of oxygen and carbon dioxide in order to catalyse mineral oxidation reactions.

Most cultures used for the commercial liberation of gold from refractory ores are natural mixtures of bacterial types. Such mixtures exhibit synergic interactions which enhance the growth of their constituents and the kinetics of bacterial oxidation reactions.

The mesophiles and moderate thermophiles form the basis of the main work which has led to the testing and commissioning of pilot and full-scale plants for the bacterial oxidation of refractory minerals. The important bacteria and their properties are described fully in Chapter 3.

Fig. 1.1 shows three examples of the enhancement of recovery by cyanidation of gold from refractory concentrates from different parts of the world. The results were derived from laboratory testing of the concentrates and, in the case of the Australian concentrate, have been confirmed at the pilot plant scale.

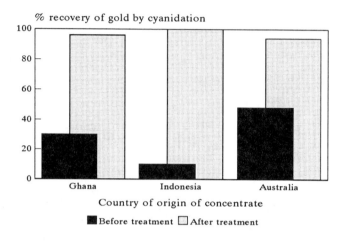

Fig. 1.1 Examples of the enhancement of gold recovery resulting from the bacterial oxidation treatment of three refractory concentrates

All three concentrates contained arsenic (the Australian samples containing as much as fourteen percent (w/w) of the element) and the Ghanaian sample contained five percent of carbon. The figures for gold recovery before bacterial oxidation treatment indicate the fraction of free gold present. After treatment the recovery was between ninety-five and one hundred percent. The carbon in the Ghanaian sample was not activated and did not consume extra cyanide solution, as can be the case in other processes.

1.8 OUTLINE OF THE BOOK

This book consists of a discussion of potentially treatable minerals (Chapter 2), a description of the bacteria which may be used as catalysts (Chapter 3) and a detailed discussion of the chemistry of the bacterial oxidation of sulfidic minerals (Chapter 4). A discussion, based upon published experimental data, of the possible mechanism of bacterial oxidation follows (Chapter 5). The designs and operation of pilot and full-scale plants are then described and related to the perceived understanding of the process (Chapter 6). This is followed by a discussion of the problems associated with the disposal and environmentally acceptable storage of the waste products (Chapter 7). The economics of the process are then discussed (Chapter 8) and finally there are details of the essential analytical methods which are used to monitor the bacterial oxidation process (Chapter 9).

1.9 REFERENCES

Brierley, C.L. (1974) *J. Less Common Metals* **36**, 237.
Brierley, C.L. & Brierley, J.A. (1973) *Can. J. Microbiol.*, **19**, 183.
Brierley, J.A. & Le Roux, N.W. (1977) *Bacterial Leaching*. Schwartz, W. (ed.) Verlag Chemie, p.55.
Brierley, J.A., Norris, P.R., Kelly, D.P. & Le Roux, N.W. (1978) *Eur. J. Appl. Microbiol. Biotechnol.*, **5**, 291.
Carvalho, T.M., Haines, A.K., da Silva, R.E.J. & Doyle, B.N. (1988) *Randol International Gold Conference, Perth, 1988*, p.152.
Colmer, A.R. & Hinkle, M.E. (1947) *Science* **106**, 253.
Ehrlich, H.L. & Brierley, C.L. (eds) (1990) *Microbial Mineral Recovery*. McGraw-Hill.
Golovacheva, R.S. & Karavaiko, G.I. (1978) *Mikrobiologiya* **47**, 5815.
Hughes, M.N. & Poole, R.K. (1989) *Metals & Micro-organisms*. Chapman & Hall.
Hutchins, S.R., Davidson, M.S., Brierley, J.A. & Brierley, C.L. (1986) *Ann. Rev. Microbiol.*, **40**, 311.
Karavaiko, G.I. & Groudev, S.N. (eds) (1985) *Biogeotechnology of Metals, Proceedings of an International Seminar on Modern Aspects of Microbiological Hydrometallurgy and International Training Course on Microbiological Leaching of Metals from Ores, Moscow, Sophia, 1982*. Centre for International Projects GKNT, Moscow.
Karavaiko, G.I., Rossi, G., Agate, A.D., Groudev, S.N. & Avakyan, Z.A. (eds) (1988) *Biogeotechnology of Metals*. Centre for International Projects GKNT, Moscow.
Karavaiko, G.I., Rossi, G. & Avakyan, Z.A. (eds) (1990) *Proceedings of an International Seminar on Dump and Underground Bacterial Leaching of Metals from Ores, Leningrad, 1987*. Centre for International Projects, USSR State Committee for Environmental Protection, Moscow.
Le Roux, N.W. & Marshall, V. (1977) *Bacterial Leaching*. Schwartz, W. (ed.) Verlag Chemie, p.21.
Matthews, D.C. (1990) *Randol Gold Forum, Squaw Valley, 1990*, p.143.
Nobar, A.M., Ewart, D.K., Al Saffar, L., Barrett, J., Hughes, M.N. & Poole, R.K. (1988) *Biohydrometallurgy - 87, Warwick, 1987, Science and Technology Letters*
P.R.Norris, P.R. & Kelly, D.P. (eds) p.530.
Rossi, G. (1990) *Biohydrometallurgy*. McGraw-Hill.
Schlegel, H.G. (1986) *General Microbiology*. 6th edn, Cambridge University Press.
Silverman, M.P. & Lundgren, D.G. (1959) *J. Bacteriol.*, **77**, 642.

2

Potentially treatable minerals

2.1 INTRODUCTION

Whether a particular mineral or mixture of minerals could be, or should be, treated by using bacterial oxidation technology depends upon a combination of scientific and economic factors. This chapter consists of an outline discussion of the various factors which should be considered in order to decide upon the feasibility of the technology for any given case.

The treatment of a mineral by bacterial oxidation reactions depends upon the chemical characteristics of the mineral, together with the grade and value of its metal content. There are three classes of bacterial oxidation reactions which may be carried out by one or other of five methods.

2.1.1 Classification of bacterial oxidation reactions

Chemolithoautotrophic bacteria require their substrates to contain iron(II) or reduced sulfur (including sulfides, disulfides or arsenosulfides) or both, depending upon the bacterial species, in order to gain energy for cell maintenance and growth. Such requirements determine the nature of the minerals which may be primarily solubilized by bacterial oxidation. In addition the products of the oxidation (sulfuric acid and/or iron(III)) of suitable minerals may be utilized for the solubilization of secondary minerals present in the ore or concentrate. The treatment of a mineral which contains neither iron(II) nor reduced sulfur is thus possible if that mineral is capable of being solubilized by acid or by iron(III) or by a combination of both of these reactants. This factor extends the number and range of minerals which are treatable by bacterial oxidation. In such cases it is necessary to have present, or to add, a sufficient pyrite (or an alternative iron(II)/sulfur-containing mineral) content.

Such considerations allow bacterial oxidation processes to be divided into three classes, which may be termed metal liberation, and primary and secondary mineral oxidations, and are described briefly as follows.

Class I: Metal liberation

In these processes the various minerals which encapsulate elementary gold (and silver) particles are oxidized and solubilized causing the liberation of the valuable metal content.

Class II: Primary mineral oxidation

This class consists of those processes in which sulfide minerals are oxidized and solubilized (or converted into insoluble sulfates) to allow the recovery of their metal content. The reaction rates are usually enhanced by the participation of the iron(III) and sulfuric acid produced by the bacterial oxidation of any pyrite content which is either naturally present or added deliberately. The enhancement of the overall rates of bacterially catalysed oxidation reactions caused by the presence of iron(III) is because of the direct involvement of the iron(III) in the primary oxidation process. It is not due to a purely chemical interaction (see section 5.3.3.1).

Class III: Secondary mineral solubilization

This class consists of the processes in which secondary minerals (oxides and carbonates) are solubilized. Secondary minerals are defined, for biohydrometallurgical purposes, as those which contain the metal values but are not capable of participating in primary bacterial oxidation because they contain neither iron(II) nor reduced sulfur. Their metal values may be recovered by allowing the primary oxidation of pyrite, or similar iron/sulfur minerals, to provide iron(III) and sulfuric acid solutions which solubilize the metal content. The solubilization reaction may also involve the oxidation of the metal to a higher oxidation state than it possessed in the mineral. It is possible in any particular operation (involving a mixture of minerals) for any two or even all three of these classes to contribute to metal recovery.

2.1.2 Methods of carrying out bacterial oxidation reactions

There are, in general, five methods of applying bacterial oxidation technology to suitably amenable substrates. The grade of the ore to be treated and whether a suitable concentrate can be produced, together with the unit price of the metal, determine the method which should be adopted. The methods are described briefly in this section and are dealt with fully in Chapter 6.

(i) Agitated tank (or stirred reactor) leaching

This involves the agitation, by stirring or air-blowing, of a slurry of the target material in an acid-resistant tank. Agitated tank leaching would be chosen for the metals with a high unit price since it is the most expensive method. Only in the case of gold, with its high price, would it be an economic possibility to use agitated tank leaching for an ore, rather than a concentrate.

(ii) Heap leaching

This involves a heap of ore which has been specifically designed and built to optimize bacterial oxidation reactions. This method can be carried out with ore or with flotation tailings, there being costs associated with the comminution of the ore or the agglomeration of the tailings to achieve the optimum particle size distribution. It is less expensive than tank leaching and can be carried out for low grade gold ores and those metals with relatively low unit prices. Without much consideration for the particle size distribution, heap leaching of copper has been shown to be very successful and so the method should be successful with suitable minerals containing metals with a higher unit price than that of copper.

(iii) Vat leaching

This method could be described as the reverse of heap leaching, in that the leach solution is allowed to percolate upwards through the pile of material as the vat is flooded with water (or preferably acid mine drainage). Vat leaching is operated on a lower scale than heap leaching. It has, to date, only been used for copper solubilization without regard to optimizing the bacterial oxidation process. It would be appropriate for the treatment of relatively small quantities of reasonably high grade ores.

(iv) Dump leaching

This method is similar to heap leaching except that a dump is a pile of material which has not normally been constructed with any purpose other than to store either mine waste material which includes untreated low grade (below cut-off grade) ore or tailings from previous extraction processes. The method allows very low grade material to be treated and, as in the case of heap leaching, it is successful for copper recovery. Without being designed for the purpose, dumped sulfidic materials undergo bacterial oxidation with the production of acid mine drainage. Much research is currently being carried out with the aim of preventing such bacterial activity. The prevention and control of acid mine drainage is crucial to the protection of the environment. The widespread occurrence of acid mine drainage is an indication of the efficiency and facility by which bacterial oxidation happens without any interference from man.

(v) *In situ* (or in-place) leaching

This method makes use of cracks and fissures, which may be natural or caused by blasting, in an ore body for the passage of oxygenated water. Mineralized agglomerates which are permeable by water also represent suitable candidates for *in situ* leaching. *In situ* leaching of copper and uranium are established procedures. The copper leaching depends upon primary bacterial action on the copper sulfide minerals coupled with the primary oxidation of iron/sulfur minerals to give the iron(III) and sulfuric acid which are needed to enhance the overall process (class II). The uranium leaching is entirely dependent upon the primary bacterial oxidation of pyrite minerals to provide the iron(III) and sulfuric acid which are needed to dissolve the uranium oxide ore (a class III process). Although the unit price of uranium is high the ore grades are usually low and tank leaching would require a concentrate to be produced. The *in situ* process is only applicable to those elements which are solubilized by the bacterial oxidation process and so is precluded for the liberation of elemental gold and silver. Any tin, antimony or lead present in the ore body of an *in situ* operation would remain in the ground as their sulfates are insoluble.

In general the metals with lower unit prices, although being recoverable by a tank leach, would normally be produced by that method only as by-products of the recovery of a metal (such as gold) with a much higher unit price. The metallurgical industry should consider the use of bacterial oxidation for the lower priced metals as an alternative to the current roasting methods. The detailed economics are dealt with in Chapter 8 but the general conclusion is that bacterial oxidation, using the agitated tank method, is associated with running costs which are slightly more than those of the roasting method but has very considerably smaller capital costs.

A general block diagram summarizing the methods of carrying out bacterial oxidation reactions and their associated processes is shown in Fig. 2.1. Detailed descriptions and discussions of the various methods are to be found in Chapters 6 (plant applications) and 8 (economic factors). Whether a particular potentially treatable mineral should be subjected to bacterial oxidation is dependent upon the economic considerations of the process as compared to those of currently used methods. The detailed consideration of the economics of bacterial oxidation processes form the basis of Chapter 8.

The following section consists of a general discussion of a selection of minerals whose methods of treatment belong to one or other of the classes described above. The examples chosen are based upon the world price and world production figures for 1990 as extracted from the 1991 Annual Review of Metals and Minerals (Volume 2 of the Mining Annual Review published by the Mining Journal, 1991). The feasibility of the adoption of bacterial oxidation technology in any particular case depends upon the suitability of the mineral as a bacterial substrate, the unit price of the metal to be extracted and the value of the annual production. The choice of method of bacterial oxidation suitable for each element is tentatively indicated. The actual choice, for any particular ore or concentrate, would depend upon the results of extensive amenability testing. At the end of the chapter there is a discussion of the geochemical and chemical classifications of minerals and their relevance to the indication of which minerals should be capable of primary oxidation.

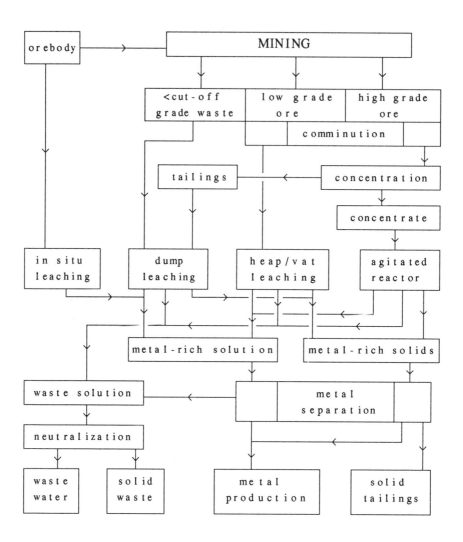

Fig 2.1 A block diagram of possible bacterial oxidation methods

2.2 WORLD PRODUCTION OF SOME METALS

The potential of bacterial oxidation technology may be put in perspective by a consideration of the value of the metals which can be solubilized or liberated by its application. The annual production (in kilotonnes per annum), unit prices (in US $ per tonne) and total annual value (in US $M per annum) for eleven metals are shown in Table 2.1. The figures are based upon the exchange rate of US $1.70 = £1. Most of the figures do not include the output information from the ex-communist countries which is generally unavailable. When such information does become available it will have a considerable effect upon some of the overall numbers.

Table 2.1 World production, prices and annual values of some metals in 1990

metal	production kt/year	price $/t	value $M/year
gold	1.73	11.63M	20192
silver	11.4	131.5k	1497
cobalt	20.9	24.6k	513
uranium	28.6	20.8k	593
nickel	930	7383	6863
molybdenum	108	6000	648
tin	158	5530	874
copper	7283	2338	17024
antimony	107	1650	176
zinc	5400	1003	5400
lead	2319	510	1180

The metals listed in Table 2.1 are in descending order of their unit prices and are considered separately in that sequence in the following sub-sections.

2.2.1 Gold

Gold is the highest priced metal, its production having the highest annual value. The distribution of world production of the metal is shown in Table 2.2.

Table 2.2 Distribution of world gold production (1990 figures)

Country	production %
South Africa	34.9
USA	17.0
Australia	13.9
Canada	9.5
Brazil	4.5
Philippines	2.1
Others	18.0

The element exists in nature mainly as the native metal, there being a small fraction of gold tellurides in some rare deposits. Between seventy to eighty-five percent of native gold is free-milling and is easily extracted from run-of-mine material by the conventional methods of comminution followed by gravity separation and/or cyanidation. Problems arise with the remaining fifteen to thirty percent of gold which is rendered refractory by being encapsulated by minerals, the more important of these being arsenopyrite, pyrite, pyrrhotite and chalcopyrite. The mineral coating is impervious to aqueous cyanide ions and drastically reduces the efficiency of recovery of the gold content by direct cyanidation. The metal in any particle is only dissolved if part of it is exposed, and therefore available, to the cyanide solution. Pre-treatment of such refractory gold-bearing minerals by the conventional roasting, or the newer pressure oxidation or bacterial oxidation, technologies liberates the gold metal fully and allows its recovery by cyanidation.

Whatever method of pre-treatment is adopted it is usually economically desirable to obtain a concentrate by flotation. Bacterial oxidation of a gold-bearing ore has been attempted at the Tonkin Springs operation (Foo *et al.* 1990), but the mechanical agitation of a material which consists of around ninety-eight percent of gangue contributes negatively to the process economics. Some refractory gold minerals cannot be concentrated by flotation (as was the case with the Tonkin Springs material). The reason for this is that the sulfide particles are physically intergrown with the gangue material on a microscopic scale and so cannot be separated by flotation even after comminution to a small particle size. In some cases partial pre-oxidation by bacterial methods renders sulfidic minerals flotable but this is, as yet, an area which has received very little investigation.

The main minerals which render gold refractory have been mentioned above and are dealt with in more detail below. All the processes which entail the liberation of gold from refractory minerals are examples of Class I.

2.2.1.1 Arsenopyrite

Arsenopyrite is the chief source of elemental arsenic. It is found as steel-grey orthorhombic crystals (Fleischer, 1987). It is often associated with pyrite and other sulfides. Its idealized formula is FeAsS but the mineral is usually non-stoichiometric. Its crystal structure (Wells, 1975) is related to that of the marcasite form of FeS_2 with half of the sulfur atoms being replaced regularly by arsenic atoms.

As arsenopyrite (and all the other sulfur-containing minerals considered in this chapter) is a semi-conductor, which implies long range interaction between the component atoms, it is an oversimplification to use the oxidation state concept to describe the electronic configurations of the constituent atoms. When subjected to bacterial oxidation the mineral is oxidized to iron(III), arsenic(V) and sulfur(VI). This process is discussed at length in Chapters 4 and 5. In general the liberation of the gold content is achieved by a seventy to eighty percent oxidation of the arsenopyrite. A typical plot of the percentage of gold recovery versus the percentage of arsenic solubilized is shown in Fig. 2.2. The results are those obtained (Spencer *et al.* 1991) by the bacterial oxidation of a sample of gold-bearing arsenopyrite/pyrite concentrate. The results obtained from laboratory batch testing on small samples (150 g) are shown in the diagram together with some results from a pilot plant running at a pulp density of fifteen percent (w/v) with a throughput of one tonne per day. The nature of the relationship between the percentage of arsenic solubilized to achieve a given percentage gold recovery depends upon the percentage of free gold in the sample, the particle size of the

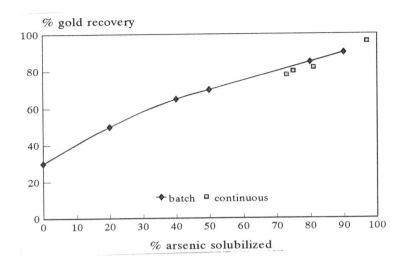

Fig. 2.2 Variation of percentage gold recovery with the percentage of arsenic solubilized in the bacterial oxidation of a refractory gold-bearing arsenopyrite/pyrite concentrate

encapsulated gold and upon the distribution of the gold throughout the sample. In the limit of the gold being so finely distributed as to be regarded as being in solid solution the solubilization of the refractory material would have to be complete to achieve the complete liberation of the metal content.

2.2.1.2 Pyrite

Pyrite (iron pyrites, FeS_2) is the most widespread and abundant sulfur mineral, and is found as either well-formed cubic (sometimes octahedral) crystals or in a framboidal (raspberry-like) massive form. It has a characteristic yellow metallic lustre. Although it is sometimes mistaken for elemental gold it has a more yellow, brass-like, appearance than the precious metal and is considerably harder. The solid may be regarded formally as containing Fe^{2+} and S_2^{2-} ions with the sodium chloride structure (Wells, 1975). There is no doubt about the atomic arrangement but, since the material is a semi-conductor, the assignments of charges or oxidation states are oversimplifications of the nature of the bonding.

2.2.1.3 Marcasite

Marcasite, FeS_2, is dimorphic with pyrite and has a crystal structure related to that of rutile (Wells, 1975). It is by no means as abundant as the pyrite form and exists in many crystalline forms. Its metallic lustre is of a somewhat paler yellow colour than that of pyrite. It tends to be more easily oxidized than pyrite. It is included in this chapter because of the relationship between its structure and that of arsenopyrite.

2.2.1.4 Pyrrhotite

Pyrrhotite (sometimes referred to as magnetic pyrites) is found as monoclinic or hexagonal crystals which are a dullish yellow colour. The crystals tarnish easily to a rusty brown colour when exposed to the air. This property is consistent with the relatively high rate at which pyrrhotite undergoes bacterial oxidation.

Although regarded as a form of iron(II) sulfide, FeS, pyrrhotite has a non-stoichiometric defect structure in which up to one eighth of the metal sites are vacant (Wells, 1975). For this reason the formula is sometimes written as Fe_7S_8, although the iron content may be more than so indicated in some samples. The mineral has a crystal structure based upon that of nickel arsenide (Wells, 1975) with up to one eighth of the iron sites being vacant. The material is magnetic and is a semi-conductor. In simple terms it may be thought of as containing iron(II) and monosulfide ions which are oxidized to iron(III) and sulfur(VI) respectively when subjected to bacterial oxidation.

It has been reported that pyrrhotite is the main source of the world's nickel in that some of the iron sites in the lattice may be replaced by nickel atoms. This is a possibility but the confusion arises because the mineral pentlandite (a mixed iron nickel sulfide, usually formulated as $(Fe,Ni)_9S_8$, discussed in section 2.2.5), which has a crystal structure very different from that of pyrrhotite, is often found intergrown with pyrrhotite (Boldt, 1966a). There is a series of iron sulfide minerals of general formula, Fe_nS_m. If n is less than m the minerals are pyrrhotites with crystal structures which may be described in terms of superstructures of the defect nickel arsenide arrangement. If $n = m = 1$ the mineral is known as troilite and has a structure similar to that of lithium hydroxide (Wells, 1975). If n is greater than m the crystal structure changes to that of pentlandite and it is in this mineral in which iron and nickel may co-exist. The iron:nickel ratio in pentlandite is fairly close to unity.

2.2.1.5 Chalcopyrite

Chalcopyrite, $FeCuS_2$, exists as tetragonal crystals with a metallic lustre which is very similar to that of pyrite, the colour being a slightly deeper yellow. In its massive form it is less lustrous than pyrite and sometimes has a characteristic iridescence. Its crystal structure is best regarded as being a zinc blende arrangement in which the zinc atoms are replaced in a regular manner by iron and copper atoms in a 1:1 ratio (Wells, 1975). The mineral is a semi-conductor and there are somewhat superfluous arguments about whether the oxidation states of the iron and copper are respectively (II) and (II) or (III) and (I). Primary bacterial oxidation results in the solubilization of the iron and copper in their (II) states. In a secondary process the iron(II) is oxidized to iron(III). The sulfur is present as the monatomic sulfide ion rather than the disulfide ion which is typical of most minerals termed as pyrites.

2.2.1.6 Relative rates of liberation of gold from refractory minerals

The minerals which render gold refractory undergo bacterial oxidation at different rates. Because of the broad range of rates at which different samples of minerals react (Groudev, 1985) it is not possible to give a definite order of reaction. The best generalization is that the order of decreasing rate of bacterial oxidation is: pyrrhotite > arsenopyrite > pyrite > chalcopyrite, dependent upon the individual samples. No comparative data are available for marcasite, but its oxidation rate is normally higher than that of pyrite.

Chemical oxidation (i.e. in the absence of bacteria) of these minerals can be carried out using either acid or an acidic solution of iron(III). In almost all cases atmospheric oxygen was not eliminated and could very well be the main oxidant with the acid and/or iron(III) acting catalytically. One case has been reported (Ichikuni, 1960) in which oxygen was excluded from an attempted oxidation of chalcopyrite with 0.1 M hydrochloric acid. The rate in the absence of oxygen was minimal even after four hours treatment at 80-85°C, the copper(II) concentration being only 0.1 mg L^{-1} or 1.57 µM. Under the same conditions, but in the presence of oxygen, the copper(II) concentration rose to 150 mg L^{-1} (2.36 mM) in the same time. Iron(III) leaching seems to be effective only at elevated temperatures. At 30°C less than five percent of chalcopyrite is solubilized after two hours and most of the reaction occurs in the first twenty minutes. Approximately complete solubilization occurs after about eighty minutes if the solution is maintained at 100°C. In this work there was no attempt to investigate the effect of eliminating the atmospheric oxygen.

Where published comparisons are available (Groudev, 1985) it appears that bacterially catalysed oxidations of pyrite, arsenopyrite and chalcopyrite are faster than abiotic chemical attacks under equivalent conditions by factors of forty seven, twelve and twenty, respectively.

2.2.1.7 Methods of bacterial oxidation of gold-bearing minerals

The choice of method of bacterial oxidation for any particular material depends upon its gold grade. For a flotation concentrate containing typically between thirty and one hundred grams of gold per tonne the agitated tank reactor method is normally chosen. The use of power for stirring, pumping, air-blowing and heating (and cooling in some variations of the technology), together with the capital expense involved in building plant, contribute to costs which are small in relation to the value of the gold released (see Chapter 8).

Where a concentrate cannot be produced there is a difficult decision to be made between tank leaching and one or other of the static methods. Only if the gold grade is relatively high does the tank leach method approach reasonable economics. For low grade material, heap or vat leaching methods are possible although they both suffer from the disadvantage that the solid product, containing the liberated gold, is highly acidic. Before extraction of the liberated gold can take place by cyanidation the material must be washed and neutralized.

In all gold producing areas of the world there are large amounts of dumped waste material and tailings from previous attempts at gold recovery. Both of these categories of materials contain low grades of gold and are potentially treatable by bacterial oxidation in reconstructed heaps. Dump leaching is probably ruled out on economic and technical grounds. The *in situ* process is inapplicable to gold recovery as it would leave the metal value in the ground.

2.2.2 Silver

The distribution of world silver production in 1990 is shown in Table 2.3.

Table 2.3 Distribution of world silver production (1990 figures)

Country	production %
USA	18.2
Mexico	17.5
Peru	15.0
Canada	12.1
Australia	10.3
Others	26.8

Whenever elemental silver co-exists with gold the two metals are extracted together by the cyanidation method. Some deposits of the native metal occur but the main supply of silver originates in sulfide minerals which are found in association with sulfides of lead, zinc and copper, and from which the metal follows the path of the base metal through its concentration and smelting processes. The silver metal is separated from the base metal in the final stages of the latter's production. The most important silver mineral is argentite, Ag_2S. Bacterial oxidation of refractory minerals encapsulating elemental silver and gold mixtures results in the liberation of any native silver associated with the gold (a class I process). As such the element is amenable to the same treatment as that given to refractory gold. There are problems associated with the treatment of silver sulfide by bacterial oxidation (a class II process). The silver sulfide would be oxidized to silver sulfate which is sparingly soluble in water (giving a solution which is 8 g L^{-1} or 26 mM in silver ion at 298K). If chloride ion is present in the system (either in the nutrient medium or the process water) the solubility of the silver ion could be much lower, the solubility product for silver chloride being 1.8×10^{-10} at 298 K. The aqueous silver ion is extremely toxic to bacteria in general (Trevors, 1987), *Thiobacillus ferrooxidans* being adversely affected by concentrations as low as 10 μg L^{-1} (79 nM) (Hoffman and Hendrix, 1976). The adaptation of chemolithoautotrophs to the resistance of silver ion concentrations of up to 0.5 mM is possible but that is still very low to be of practical use. For the present, it would seem that bacterial oxidation is unlikely to be of any value for the treatment of silver sulfide ores and concentrates. This preliminary conclusion has implications for the treatment of any mixed sulfidic material which has a significant silver sulfide content. In these cases it would be necessary to ensure that sufficient chloride ion was present to suppress the solubility of the silver ion.

2.2.3 Cobalt

Cobalt is found mainly as the sulfide/arsenosulfide minerals cobaltite, CoAsS, and linnaeite, Co_3S_4, and as the arsenide known as smaltite, $CoAs_2$. The structure of cobaltite (Wells, 1975) is based upon that of pyrite with one half of the S-S groups being replaced by As-As groups. Linnaeite has a spinel structure in which two thirds of the cobalt atoms are in oxidation state (III), the other third being in oxidation state (II). Half of the octahedral sites and one eighth of the tetrahedral sites in the cubic close packed sulfide lattice are filled. Smaltite is a non-stoichiometric arsenic-deficient form of the mineral skutterudite, $CoAs_3$. The cobalt sulfides are usually found as minority components of copper, nickel and lead sulfide minerals, the metal being extracted chemically from their roasting products. The bulk of the world's production of cobalt comes from central Africa as is shown in Table 2.4.

Solubilization of the cobalt content of cobalt sulfide/arsenosulfide minerals as the aqueous Co(II) ion may be brought about by the action of sulfide oxidizing bacteria (Torma, 1971), particularly when iron is present in the system (a class II process). It would seem that bacterial oxidation would be a viable treatment for these ores, in concentrate form, as an alternative to current smelting methods. It is probable that the agitated tank reactor method would be appropriate for the relatively high valued product.

Table 2.4 Distribution of world cobalt production (1990 figures)

Country	production %
Zaire	48.1
Zambia	23.2
Canada	18.7
Finland	6.2
Others	3.8

Any cobalt sulfides present in any of the four static leaching operations for the solubilization of major metals such as copper would lead to the release of the cobalt content as aqueous Co(II) ions.

Smaltite is not a suitable substrate for any primary bacterial oxidation process but would be completely solubilized as a secondary mineral if added to a mixture of suitable iron/sulfur minerals (a class III process).

2.2.4 Uranium

The distribution of world uranium production in 1990 is shown in Table 2.5.

Table 2.5 Distribution of world uranium production (1990 figures)

Country	production %
Canada	30.4
USA	12.4
Australia	12.3
Europe	11.3
Namibia	11.2
Niger	9.9
South Africa	8.9
Others	3.6

Uranium exists as the oxides uraninite (pitchblende), UO_2, and U_3O_8, the mixed oxides brannerite, UTi_2O_6, and coffinite, $USiO_4.xH_2O$, and some minerals which contain the uranyl cation, UO_2^{2+}; carnotite, $K_2(UO_2)_2(VO_4)_2.3H_2O$, and autonite, $Ca(UO_2)_2(PO_4)_2.10H_2O$. The minerals which contain the uranyl cation dissolve easily in dilute sulfuric acid. It is necessary that the uranium in the oxide minerals is oxidized to the (VI) state (UO_2^{2+}) to achieve solubilization. The solubilization of uranium by bacterial oxidation depends upon the presence of associated pyrite and is an excellent example of the solubilization of a metal from ores which themselves have no content which is of interest to iron/sulfide oxidizing bacteria (a class III process).

The bacterial action upon any pyrite content of the uranium ore produces sulfuric acid and iron(III), the former solubilizing the minerals containing uranyl ions and the latter oxidizing uranium oxides (Tuovinen et al. 1983). The oxidation and solubilization of uranium(IV) oxide may be written as the equation:

$$2Fe^{3+} + UO_2 \rightarrow UO_2^{2+} + 2Fe^{2+} \tag{2.1}$$

The iron(II) produced is re-oxidized to iron(III) by the bacteria which use oxygen as the oxidant.

Although several pilot plants have been designed (e.g. Derry et al. 1977) and have been used for uranium solubilization there have been no reports of any full scale plants. The *in situ* solubilization of uranium at the Elliot Lake and Agnew Lake mining areas in Canada has been reported in some detail (McCready, 1988, McCready and Gould, 1989). The technique developed involved the use of worked out stopes in the underground mines. The roof of the stope was blasted so that the roof material (containing uranium minerals) was reduced to appropriately sized particles (two to twenty centimetres diameter). The stope was blocked off by a reinforced concrete bulkhead and was then flooded with acid mine drainage solution which had a pH value of 2.3. The flooding took three days after which the solution was drained out. The drained out stope was allowed to react for three weeks before the flooding/drainage cycle was repeated. This is effectively a mixed *in situ*/vat leaching process, making use of the indigenous *T. ferrooxidans*, and was found to be very effective with the added advantage of not producing any tailings, the leached material remaining underground. McCready and Gould (1989) report that the bioleaching of uranium is substantially cheaper than the conventional sulfuric acid extraction process. Uranium(VI) is toxic to unadapted *T. ferrooxidans* cultures at concentrations greater than 0.7-12 g L^{-1} (3-50 mM) (Duncan and Bruynesteyn, 1971, Ebner and Schwartz, 1974). Without bacterial adaptation to higher values such data imply limitations for the use of stirred reactors for uranium extraction. It would be expected that low grade ores in heaps, vats or dumps would yield to bacterial oxidation treatment, although there are no reported attempts to date.

There are some *in situ* uranium mining operations which are established and which do not depend upon bacteria. A description of one such process is given by Hunter (1991) in which groundwater, enriched with oxygen and carbon dioxide, is pumped down boreholes into mineralized sandstone conglomerate beds. Oxygen oxidizes the uranium to its (VI) state in which it is stabilized as carbonate complexes. Solution containing the solubilized uranium is collected from other boreholes to be treated by ion exchange to concentrate the metal value.

2.2.5 Nickel

About seventy-five percent of nickel production is from the iron/nickel sulfide pentlandite, $(Fe,Ni)_9S_8$. Minor sources include violerite, $(Ni,Fe)_3S_4$, bravoite, $(Ni,Fe)S_2$, nickeline (or niccolite), NiS, millerite, NiS, ullmannite, NiSbS, the silicate, garnierite, $(Ni,Mg)_6Si_4O_{10}(OH)_8$, and nickeliferous limonite, $(Fe,Ni)O(OH).nH_2O$, the latter two minerals being constituents of laterite. Niccolite is mentioned because it possesses the nickel arsenide structure in which the nickel atoms occupy one half of the tetrahedral holes in a hexagonally close packed lattice of sulfur atoms (Wells, 1975). The importance of the nickel arsenide structure is discussed below (section 2.4.2). Millerite has the same stoichiometry as niccolite but has a 5:5 coordination in its lattice.

The structure of pentlandite is of considerable interest. It is based upon the cubic close packing of the sulfur atoms. Sets of fourteen sulfur atoms forming cubes (one atom per corner and one atom in each face centre) are filled alternately with either eight metal atoms which occupy tetrahedral holes (either iron or nickel in varying proportions) or with one metal atom in the cube centre which is an octahedral hole. The cube containing the eight metal atoms has the anti-fluorite structure exemplified by lithium monoxide, Li_2O. The clusters containing eight metal atoms allow metal-metal bonding (Burdett and Miller, 1987) to occur, the metal-metal distances being much the same as in the pure metals. It is probable that the metal-metal bonding is the stability factor which makes pentlandite the predominant nickel-sulfur mineral.

The distribution of world production of the metal in 1990 is shown in Table 2.6.

Table 2.6 Distribution of world nickel production (1990 figures)

Country/industry	production %
INCO (Canada, Indonesia)	19.3
Other Western sources	42.4
USSR	32.5
Cuba	2.9
China	2.9

The main other western sources are the USA, N. Caledonia, W. Australia, the Philippines and Japan. Conventional extraction (Boldt, 1966b) consists of roasting with added silica to remove the iron sulfides, followed by a smelting stage to produce the metal. The nickel sulfide minerals containing iron are most appropriate for bacterial oxidation (a class II process). The sulfur content provides sulfuric acid, the iron being solubilized as iron(III) which aids the sulfide oxidation. Other nickel minerals which could be bacterially oxidized are millerite and ullmannite but they would be best accompanied by added pyrite. The silicate minerals would not be treatable by bacterial oxidation methods.

The treatment of the appropriate minerals by bacterial oxidation would produce the aqueous nickel(II) ion. This would then have to be separated from the iron(III) by a selective precipitation process, or by the use of electrowinning or liquid ion exchange, to recover the pure metal. Good recoveries have been reported (Torma, 1971) although the method is no better than the use of a sterile acidic iron(III) solution (Dutrizac and MacDonald, 1974). This is not surprising since an acid/iron(III) solution is all that is needed to dissolve nickel sulfide minerals. The advantage of bacterial oxidation is that the acid/iron(III) mixture is produced in the process and does not have to be bought in to carry out the solubilization.

Although no pilot plant scale work has yet taken place on the bacterial solubilization of nickel there seems to be every possibility that it would be an efficient and economic method for recovering the element. The unit price of nickel is considerably lower than those of the previous elements considered. It is probable that tank leaching would be economically viable although some form of heap or vat leaching would possibly be a more appropriate method.

2.2.6 Molybdenum

Molybdenum occurs mainly as molybdenite, MoS_2, which is a by-product of copper sulfide flotation concentration. Conventionally the sulfide concentrate is roasted to give the oxide, MoO_3, most of which (eighty-five percent) is then used directly in the manufacture of stainless steel. The distribution of world production in 1990 is shown in Table 2.7.

Table 2.7 Distribution of world molybdenum production (1990 figures)

Country	production %
USA	63.1
Chile	13.8
Canada	12.0
Others	11.1

The crystal structure of MoS_2 (Wells, 1975) is best regarded in terms of a hexagonally closest packed arrangement of sulfide ions in which molybdenum atoms occupy one half of the trigonal prismatic holes between the layers of sulfide atoms but only make use of alternate layers. The molybdenum is in its formal (IV) oxidation state. There is evidence (Bryner and Anderson, 1957) that bacterial oxidation can solubilize molybdenum in the absence of pyrite, but the presence of the iron mineral boosts the oxidation rate by a factor of three (a class II process). The molybdenum is solubilized in the (VI) state as the green ion, MoO_2^{2+}, in the strongly acidic conditions at which bacterial oxidation occurs. At higher pH values (up to four) Mo(VI) occurs as the neutral molybdic(VI) acid, H_2MoO_4 (this is the simplest form of the acid, but it exists in a number of polymeric forms).

It has yet to be shown that catalytic bacteria can resist the toxic effect of solubilized molybdenum. Concentrations between five and ninety ppm (0.05 - 0.1 mM) have been reported to be toxic (Tuovinen et al. 1971), but it is possible that suitably chosen cultures would adapt to higher tolerances. It is possible that the toxicity of Mo(VI) would be very dependent upon the pH value of the leach solution. The neutral molybdic acid would be expected to be much more toxic than the green positive ion mentioned above. C.L. Brierley (1974) has pointed out that the extreme thermophile, Sulfolobus, can catalyse the oxidation of molybdenite and tolerate an aqueous molybdenum(VI) concentration of up to 750 ppm (8 mM). She also shows that, when iron(III) is present, some of the molybdenum is precipitated as iron(III) molybdate.

The unit price of molybdenum is sufficiently high for tank leaching to be a viable method for the bacterial oxidation of its sulfide in the form of a concentrate. The success of the method would be dependent upon the adaptation of the chosen culture to high molybdenum(VI) concentrations. Heap or vat leaching would be possible if the concentrate were to be agglomerated to give a suitable particle size distribution.

2.2.7 Tin

The main ore of tin is the oxide cassiterite, SnO_2, from which the metal is obtained conventionally by reduction with carbon. The oxide ore is not one in which solubilization of the metal could be achieved by the bacterial oxidation of a pyrite admixture because of its slow reaction with acid. There is, however, a mixed copper/iron/tin sulfide, stannite, Cu_2FeSnS_4, which has a crystal structure (Wells, 1975) based upon that of chalcopyrite in which half of the iron atoms are replaced by tin. Although no research has been published on the bacterial oxidation of this mineral it would be expected that the tin content would be converted to the insoluble tin(IV) sulfate, $Sn(SO_4)_2$ (a class II process). The initial product of bacterial oxidation would be the soluble tin(II) ion, but this is rapidly oxidized to tin(IV) by iron(III) which would be present in the system. The copper content would be solubilized as the copper(II) ion as in the cases of the copper minerals discussed in section 2.2.8.

The fraction of total tin production which originates from the smelting of stannite is very small, but it would be beneficial for there to be small bacterial oxidation plants set up at mines which have a significant stannite production (e.g. Neves Corvo in Portugal). Stannite is usually produced by differential flotation and, as a concentrate, would have to be treated by the tank reactor method since the static methods would leave the tin in the solid state as the insoluble Sn(IV) sulfate.

The distribution of world production of tin in 1990 is shown in Table 2.8.

Table 2.8 Distribution of world tin production (1990 figures)

Country	production %
Brazil	25.8
Indonesia	20.0
Malaysia	18.7
Bolivia	11.0
Thailand	9.7
Others	14.8

2.2.8 Copper

Copper is mined on a massive scale, the main minerals being chalcopyrite, $CuFeS_2$, covellite, CuS, chalcocite, Cu_2S, bornite, Cu_5FeS_4, and enargite, Cu_3AsS_4. The element is also found as the native metal and as the basic carbonate mineral, malachite, $CuCO_3.Cu(OH)_2$. The sulfidic minerals are treated conventionally by flotation followed by roasting. The distribution of the world production of copper, by continent, in 1990 is shown in Table 2.8.

As is pointed out in Chapter 1 the bacterial oxidation of copper ores has been recognized as contributing to the production of the metal since the discovery of *Thiobacillus ferrooxidans* in 1947. The process has, of course, been occurring for as long as copper deposits have been mined and probably occurred (and is occurring) in the natural untouched deposits wherever suitable channels allowed the passage of oxygenated water. It has been estimated that bacterial leaching is responsible for at least fifteen percent (1.1 Mt) of the world production of the metal.

Table 2.8 Distribution of world copper production (1990 figures)

Continent	production %
America(N)	33.5
America(S)	30.5
Africa	16.1
Asia	7.6
Australasia	6.9
Europe	5.4

The understanding of the bacterial oxidation of copper sulfide minerals is second only to that of refractory gold-bearing minerals. Chalcopyrite has been described fully in section 2.2.1.5. Covellite (CuS), in spite of its simple formula, has a very complex structure (Wells, 1975). One third of the metal atoms (formally in the (II) state) have three sulfur neighbours at the corners of a triangle. The other two-thirds (formally in the (I) state) have four sulfur atom neighbours arranged tetrahedrally. Two-thirds of the sulfur atoms are present as the formal pyrite-type disulfide ions, S_2^{2-}, the remainder being formal monatomic sulfide di-anions, S^{2-}. Conventional oxidation state nomenclature would allow the compound to be represented by the formula, $Cu_4^I Cu_2^{II}(S_2)_2 S_2$. It has a very deep indigo-blue colour and has semi-conduction properties. Chalcocite occurs as massive black or dark grey monoclinic crystals. Some examples have copper deficiencies compared to the stoichiometric formula (Cu_2S) and are given other names (digenite and djurleite). Bornite exists as massive cubic crystals with colours ranging from copperish brown to peacock-blue. The assignments of the copper as being in the (I) state and the iron being in the (III) state are consistent with the overall formula but, as has been indicated previously, such assignments lead to an oversimplified view of the bonding in the solid. The mineral is a semi-conductor like the other sulfide minerals.

The production of solubilized copper by bacterial oxidation in agitated tanks is of doubtful economic viability because of the relatively low unit price of the element. Its production from waste dumps, tailings heaps (flotation waste) and by treating low-grade ore in vats (flooded concrete tanks) is economically viable and has been proved to be so on a large scale over many years and at many sites. Its production from dumps occurs without them being specially constructed for the purpose. Now that the solubilization of copper from these sources is known to arise by bacterial oxidation heaps can, and should, be constructed with the requirements of optimizing the bacterial action. This means constructing them with efficient means for the essential percolating water together with its dissolved oxygen and acid (if, as is beneficial, acid mine drainage water is used), and building them on suitable sloping sites which have been rendered impervious to liquid by some form of plastic sheeting. Waste heaps should be constructed from material of an optimum size distribution by extra comminution if necessary. Flotation waste may need to be agglomerated to achieve a suitably large size distribution for efficient liquid flow to occur through the heap.

There is a very useful description of *in situ* copper leaching operation (Concha *et al.* 1991) as it is carried out at the La Hermosa mine in Chile. The production of metallic copper by treating the copper sulfate effluent from the mine with scrap iron has been established for around one hundred years. The current production costs are $1285 per tonne. The cemented copper, which has a recovery efficiency of eighty-five percent, is transported by road to a smelter. The remaining solution, containing unrecovered copper(II) and iron(II) ions, is pumped to the acid make-up pond at the top of the deposit to be combined with sulfuric acid and re-cycled through the mine.

The processes described for the bacterial oxidation of copper/sulfur minerals belong to class II but any copper carbonate minerals could be treated by what would be a class III process. Any carbonate minerals co-existing with copper/sulfur mineral deposits would be solubilized in any *in situ* operations.

2.2.9 Antimony

The main mineral from which antimony is extracted is stibnite, Sb_2S_3, which exists as silver-grey crystals with a metallic lustre. Other (mixed) sulfide minerals which contribute to the world production of the metal are ullmannite, NiSbS, livingstonite, $HgSb_4S_8$, tetrahedrite, Cu_3SbS_3 (sometimes with a substantial iron content), wolfsbergite, $CuSbS_2$, and jamesonite, $FePb_4Sb_6S_{14}$. The element is produced from stibnite by reaction with metallic iron:

$$Sb_2S_3 + 3Fe \rightarrow 3FeS + 2Sb \qquad (2.2)$$

The distribution of world production of antimony in 1990 is shown in Table 2.9.

Table 2.9 Distribution of world antimony production (1990 figures)

Country	production %
South Africa	21.4
Bolivia	18.5
China	16.7
USSR	10.3
Thailand	6.0
Turkey	4.8
Mexico	3.4
Others	18.9

All the above sulfide minerals are susceptible to bacterial oxidation and admixed with pyrite would certainly be treatable by bacterial oxidation (a class II process). The unit price is fairly low and there must be some doubt as to whether tank leaching would be economically viable. With the current state of knowledge it might be best to conclude that any bacterial production of antimony should be limited to that present in small quantities in the production of other metals. Refractory gold-bearing minerals such as arsenopyrite sometimes occur in the presence of stibnite, the antimony content being solubilized together with the iron and arsenic in these cases. Antimony(III), the initially released product of the bacterial oxidation of stibnite, undergoes hydrolysis to give the antimonyl ion, SbO^+, which, in a sulfate system, is precipitated as $(SbO)_2SO_4$ (Torma and Gabra, 1977). It is for this reason that the presence of antimony poses virtually no toxicity threat to the bacteria used in the oxidation process. There is also evidence that the antimony may be oxidized as far as the (V) state, in which case the product will be $(SbO_2)_2SO_4$ which is also insoluble in water. The insolubility problems would make any of the static methods too expensive for the recovery of antimony as the major metal content of any mineral substrate.

2.2.10 Zinc

The major sources of zinc are the minerals sphalerite (zinc blende), ZnS, smithsonite, $ZnCO_3$, and calamine, $Zn_4Si_2O_7(OH)_2.H_2O$. Conventionally, the sulfide concentrate is roasted to give the oxide which is then reduced to the metal with carbon. The world distribution of zinc production in 1990 is shown in Table 2.10, by continent, with the major producer country indicated in parenthesis.

Table 2.10 Distribution of world zinc production (1990 figures)

Continent (major producer)	production %
America (Canada)	54.1 (21.9)
Europe (Spain)	17.4 (4.8)
Australia	17.1
Asia (Japan)	6.5 (2.4)
Africa (South Africa)	4.9 (1.4)

Although zinc sulfide is solubilized by bacterial oxidation its rate of solubilization is enhanced by the presence of pyrite (a class II process). The rather low unit price precludes the use of tank leaching although heap or vat leaching could very well become a feature of the extraction of zinc metal in the future. Dump and *in situ* leaching would cause any zinc content to be solubilized as the aqueous zinc(II) ion.

2.2.11 Lead

The distribution of the world production of lead in 1990 is shown in Table 2.11, by continent, with the major producer country indicated in parenthesis.

Table 2.11 Distribution of world lead production (1990 figures)

Continent (major producer)	production %
America (USA)	49.3 (21.0)
Australia	24.2
Europe (Sweden)	14.2 (3.6)
Africa (South Africa)	7.6 (3.0)
Asia (India)	4.7 (1.1)

Lead has a very low unit price which would preclude its production by tank leaching unless it was present as a minor constituent of a mineral mixture being treated for the recovery of a more valuable metal. The main mineral from which the metal is produced is galena, PbS. Conventionally this is roasted to give the oxide which is then reduced to the metal with carbon. Apart from the low unit price there is a problem in the use of bacterial oxidation concerned with the low solubility of the product, lead(II) sulfate. It has been shown that where galena is present as a minor component of bacterially oxidized mixtures of sulfides lead sulfate is amongst the products. Oxidation of the galena does take place (a class II process) and the lead sulfate is separable from the solid residue. The low solubility of the lead(II) reduces the possibility of toxicity to any bacteria. It is doubtful whether the method is applicable to major lead containing materials, but the extraction of the metal would be a bonus if it is present in any mixed sulfide mineral substrate. The selective solubilization of copper and zinc, which often occur together with galena in lead ores, could be the basis of a better process for the beneficiation of lead ores than that currently used.

2.3 CONCLUSIONS

Based upon the above discussions, the actual and potential bacterial oxidation methods which are suitable for the eleven elements considered in detail in this chapter are summarized in Table 2.12. The decisions are based upon the assumption that the element in question represents the major metal value. A '√' means that the method would be successful in recovering the metal value; it does not necessarily mean that the method would be economically viable. An '×' means that the method would not be successful for the particular reasons discussed above. The choices shown in Table 2.12 are made conservatively and refer to substrate minerals containing each particular element as the sole metal value. The conclusions may change as the results of further research are published. It is very common for any mineral deposit to contain more than one metal of value and a significant advantage of bacterial oxidation processes is that no separations of the metals need be carried out in the earlier mineral processing stages. The separation of the solubilized metals is effected on either the solid product or the effluent solution from the bacterial oxidation plant (or both) (see section 7.2).

Table 2.12 Bacterial oxidation methods suitable for the metals considered.

metal	tank	heap	dump	vat	*in situ*
gold	✓	✓	✓	✓	×
silver	×	×	×	×	×
cobalt	✓	✓	✓	✓	✓
uranium	×	✓	✓	✓	✓
nickel	✓	✓	✓	✓	✓
molybdenum	✓	✓	✓	✓	✓
tin	✓	×	×	×	×
copper	×	✓	✓	✓	✓
antimony	×	×	×	×	×
zinc	×	✓	✓	✓	✓
lead	×	×	×	×	×

There are cases where some or all of the elements, copper, antimony, zinc and lead, are present in minor quantities in tank leach operations for the recovery of more valuable components of the mineral substrate. Their minerals will be oxidized with copper and zinc being solubilized, antimony and lead being converted to their insoluble sulfates.

2.4 CLASSIFICATIONS OF POTENTIALLY TREATABLE ELEMENTS

Elements have been classified on the basis of their geochemical and chemical properties. This section consists of a comparative discussion of these classifications and is restricted to the metallic elements. It indicates the value of such classifications in deciding which elements are likely to exist in mineral forms that are suitable for treatment by bacterial oxidation methods.

2.4.1 Geochemical classification

The geochemical classification of the elements as contained by ore minerals (The Open University, 1976) consists of the two categories, (i) abundant, and (ii) scarce, based upon whether they individually constitute more than, or less than, 0.01% (100 ppm) of the Earth's crust respectively. The only abundant metal which exists partially (and substantially so) as a component of sulfide ores is iron and is dealt with in detail in this chapter. The reason for iron to be considered is because of its role, in the form of various sulfide and arsenosulfide minerals, in rendering elemental gold and silver refractory and in participating in all three classes of bacterial oxidation processes. The other abundant elements (aluminium, chromium, titanium, vanadium, manganese, magnesium, sodium and potassium) exist as either oxides, carbonates or silicates and so bacterial oxidation has no great relevance to them unless their ores are easily solubilized by acid and/or iron(III) by class III processes.

The class of scarce metals is sub-divided into three sections; (i) siderophiles, (ii) chalcophiles and (iii) lithophiles. The criterion which categorizes the scarce elements into these sections is whether they exhibit a preferential affinity for metallic iron, sulfur or oxygen/silicon dioxide, respectively. Originally these preferences were deduced mainly from the distribution of the elements found in meteorites (Goldschmidt, 1937), but the categorization is generally consistent with the relative reactivities of the metals with oxygen and sulfur. There are some borderline elements which are assigned differently in other accounts of the subject.

2.4.1.1 Siderophiles

The siderophile elements include gold, palladium, platinum, rhodium, iridium, ruthenium, osmium and rhenium. The latter element is sometimes classified as a chalcophile. They are not easily oxidized, readily alloy with iron and are likely to have been enriched preferentially in the Earth's core (considered to be principally an iron-nickel alloy). Gold is the only metal in this class to be considered in this chapter. The other elements in the class, if found as sulfides or as the metal encapsulated by sulfides, would also be amenable to extraction (Class II processes) or liberation (Class I processes) respectively by bacterial oxidation methods.

2.4.1.2 Chalcophiles

The chalcophile elements include molybdenum, cobalt, nickel, copper, silver, zinc, cadmium, mercury, lead, antimony, and bismuth. They exhibit a supposed preference to combination with sulfur (rather than oxygen) and, as it is considered that there is possibly a sulfur layer within the Earth's mantle or core, may have been formed and concentrated at some depth. If that is so then it is not surprising that they do not exist as oxides or silicates, although they do have the chemical potential so to do. The oxide of a metal of any given oxidation state is, with silver(I) being the only exception, thermodynamically more stable (i.e. has a standard Gibbs energy of formation which is more negative at 298 K) than the corresponding sulfide. This is obvious from their capacity to be roasted, by which process sulfides are converted to oxides. In the formation reaction (from the solid metal and gaseous oxygen) of an oxide there is a reduction in entropy which may be attributed mainly to the loss of the translational freedom of the gaseous oxygen molecules. The entropy change in the formation of a sulfide from its component elements (both solids) is practically zero. This implies that, as the temperature increases, an oxide becomes less stable and there may be conditions where it is less stable than the corresponding sulfide.

In the production of igneous rocks it might be expected that the chalcophiles would be associated with sulfur, because of the absence of oxygen and the high temperature. The sulfides that are commonly found in igneous rocks are pyrite and pyrrhotite, presumably because the available sulfur reacted preferentially with the relatively large amount of iron present. The chalcophile content of igneous rocks is found mainly as silicates. This is either because of the competition for the available sulfur by the abundant iron or because of the relatively low thermal stability of the chalcophile sulfides. Shales and sedimentary rocks generally have a relatively high chalcophile sulfide content which is thought to be because of their production by precipitation by hydrogen sulfide generated in marine muds. The same production method accounts for the high concentration of chalcophile elements in coal ash. Eight out of the eleven elements considered in this chapter as potentially treatable by bacterial oxidation belong to the chalcophile class.

2.4.1.3 Lithophiles

The lithophile elements include the members of groups one, two and three of the periodic classification (section 2.4.2, sodium, potassium, magnesium, and aluminium being abundant elements; they are abundant lithophiles), zirconium, hafnium, niobium, tantalum, tungsten, tin and uranium. Tin is sometimes classified as a chalcophile. They exhibit a strong affinity for oxygen and are generally found as either oxides or as silicates.

2.4.1.4 Metals recoverable by bacterial oxidation methods

The metals of greatest interest, being those recoverable by bacterial oxidation, fall almost entirely into the chalcophile category of the scarce elements apart from gold which is classified as a siderophile and uranium and tin which are lithophiles. Refractory gold is rendered so by encapsulation by iron sulfides or arsenosulfide. Iron is a very abundant metal (the second most abundant after aluminium) which is lithophilic in character, but nevertheless exists in large amounts as sulfur-containing minerals. It is probable that the origin of refractory gold is by deposition as the result of hydrothermal activity. One of the few methods of dissolving gold is to use (at high temperatures) a chloride ion solution in the presence of an oxidant. If such a solution comes into contact with bulk siderophilic gold the metal will be dissolved and re-distributed according to gravity. The dichloroaurate(I) ion, $[AuCl_2]^-$, may then be reduced to the metal (by sulfur or sulfides present), the simultaneous production of pyrite (or other sulfides or arsenosulfides etc.) from hydrothermal iron(II) ensuring its encapsulation. Although the great majority of the gold in the South African mines is generally thought to be of placer origin (produced by the erosion of bulk metal deposits) some of it is refractory. It is thought that this is due to secondary hydrothermal activity in the area.

2.4.2 Chemical classifications

The chemical properties of the elements and their compounds are usually discussed in terms of the positions of the elements in the periodic classification, a form of

1	2	3	4	5	6	7	8	9	10	11	12	13	14	15	16	17	18
1 H 2.2																	2 He
3 Li 1.0	4 Be 1.5											5 B 2.0	6 C 2.5	7 N 3.1	8 •O 3.5	9 F 4.1	10 Ne
11 Na 1.0	12 Mg 1.2											13 Al 1.5	14 Si 1.7	15 P 2.1	16 •S 2.4	17 Cl 2.8	18 Ar
19 K 0.9	20 Ca 1.0	21 Sc 1.2	22 Ti 1.3	23 V 1.5	24 Cr 1.6	25 Mn 1.6	26 •Fe 1.6	27 •Co 1.7	28 •Ni 1.8	29 •Cu 1.8	30 •Zn 1.7	31 Ga 1.8	32 Ge 2.0	33 •As 2.2	34 Se 2.5	35 Br 2.7	36 Kr
37 Rb 0.9	38 Sr 1.0	39 Y 1.1	40 Zr 1.2	41 Nb 1.2	42 •Mo 1.3	43 Tc 1.4	44 Ru 1.4	45 Rh 1.5	46 Pd 1.4	47 •Ag 1.4	48 Cd 1.5	49 In 1.5	50 •Sn 1.7	51 •Sb 1.8	52 Te 2.0	53 I 2.2	54 Xe
55 Cs 0.9	56 Ba 1.0	71 Lu 1.1	72 Hf 1.2	73 Ta 1.3	74 W 1.4	75 Re 1.5	76 Os 1.5	77 Ir 1.6	78 Pt 1.4	79 •Au 1.4	80 Hg 1.5	81 Tl 1.4	82 •Pb 1.6	83 Bi 1.7	84 Po 1.8	85 At 2.0	86 Rn
87 Fr 0.9	88 Ra 1.0	▲															
		▲▲	Transactinide elements														

▲ Lanthanide elements (La-Yb, $x = 1.0$-1.1)
▲▲ Actinide elements (including •U, $x = 1.2$)

Fig. 2.3 The periodic classification of the elements showing their Allred-Rochow electronegativity coefficients (x); elements marked with • form the main subject matter of this book

which is shown in Fig. 2.3. The basis of the classification is that elements are placed in the order of their atomic numbers, each of the vertical groups (numbered from one to eighteen in accordance with the latest IUPAC recommendations) containing elements with the same number of valency electrons. Lutetium, although still regarded as a lanthanide element according to its chemical behaviour, is shown as the first member of the third transition series. Similarly, lawrencium is shown as the first member of the fourth transition series. The chalcophile elements are found in groups 9 to 15 (with the exception of molybdenum which is in group 6), the siderophile elements being in long periods 2 and 3 (groups 7 to 11). The lithophiles are the elements of groups 1 to 7 (excluding hydrogen) plus the lanthanides (elements 57-71) and the early members of the actinides (elements 89-92, those with atomic numbers greater than 92 having no natural abundance). Aluminium, in group 13, is the most abundant metal and is lithophilic. The most useful chemical parameter, for classifying the properties of the elements with regard to their existence in nature as oxides, native metals or sulfides, is the Allred-Rochow electronegativity coefficient (Barrett, 1991). Fig. 2.3 includes the values of these coefficients. The electronegativity coefficient (x) of an element represents its tendency to attract electrons to itself in its combined state. The scale of values of x extends from 0.9 (the heavier group 1 metals) to 4.1 (fluorine). Easily oxidized elements have a low value of the coefficient, less easily oxidized elements having higher values. The elements which are very difficult to oxidize are the non-metals which have x values at the top end of the scale.

From a chemical standpoint the lithophiles are amongst the least electronegative metals (most electropositive, most easily oxidized), their electronegativity coefficients having values between 0.9 and 1.6. The chalcophiles are the more electronegative metals (less easily oxidized) with x values in the range 1.3 to 2.2. The values of x of the siderophiles lie in the range 1.4 to 1.6, intermediate between the lithophiles and chalcophiles, and are generally unreactive to either oxygen or sulfur. From a general chemical standpoint this is an indication of a kinetic inertness of the siderophile elements rather than a lack of thermodynamic stability of their oxides and sulfides. A possible reason for the distinction between the chalcophiles and siderophiles is the inability of elements of the latter class to participate in structures related to those of nickel arsenide and pentlandite (see section 2.2.5). The nickel arsenide and pentlandite structures possess metal-metal bonding interactions as evidenced by the metal-metal distances which are more or less equal to those found in the corresponding metallic structure. The siderophile elements are generally too large to fit into the tetrahedral holes in the close packed sulfur atom lattices (hexagonally close packed for nickel arsenide, cubic close packed for pentlandite) which must be occupied by metal atoms if either the nickel arsenide or pentlandite structures are to be formed.

The behaviour of metal ions and ligand groups which bind to them are subject to two chemical classifications; one being that due to Ahrland, Chatt and Davies (ACD), (1958), the other being the Hard and Soft Acid and Base (HSAB) concept

(Pearson, 1963). The ACD and HSAB classifications are both based upon observations of the behaviour of the central metal ions of complexes towards the donor atoms of the ligand groups. They are thus empirically based as is the geochemical classification.

2.4.2.1 The Ahrland-Chatt-Davies (ACD) classification

ACD '*a*-type' behaviour occurs when the orders of stability constants (formation constants) for a given oxidation state of a given element and the following series of donor ligand atoms or ions are O»S>Se>Te, N»P>As>Sb and $F^->Cl^->Br^->I^-$. The '*b*-type' behaviour occurs when the orders are O«S~Se~Te, N«P>As>Sb and $F^-<Cl^-<Br^-<I^-$. Intermediate between the *a*-type and *b*-type classes are the borderline elements. Some typical oxidation states for some elements are classified in Table 2.13.

Table 2.13 Ahrland-Chatt-Davies classifications of some metal ions (oxidation states shown by Roman numerals)

class-*a*	borderline	class-*b*
Li,Na,K(I)	Fe,Co,Ni(II)	Cu,Ag,Au,Tl(I)
Be,Mg,Ca,Sr,Sn,Mn(II)	Cu,Zn,Pb(II)	Hg,Pd,Pt(II)
Al,Sc,Ga,Sb(III)		Tl(III)
In,La,Cr,Co,Fe(III)		Pt(IV)
Ti,Sn(IV)		
Sb(V),Mo,U(VI)		

2.4.2.2 The hard and soft acids and bases (HSAB) classification

The HSAB concept is based upon the considerations that the central metal ion of a complex is an electron pair acceptor (a Lewis acid) and that the ligands are electron pair donors (Lewis bases). It applies to ligand behaviour in addition to that of the metal ions. Metal ions and the ligand atoms or ions are classified as exhibiting either 'hard' or 'soft' behaviour.

Hard acids include ions of sodium(I), potassium(I), magnesium(II), calcium(II), aluminium(III), iron(III), chromium(III), manganese(IV), tin(IV), titanium(IV), antimony(V), molybdenum (VI) and uranium(IV and VI). The hardness of the (VI) states of molybdenum and uranium are demonstrated by their bonding to the hard oxide ion in forming their MO_2^{2+} oxocations. The oxocations would be expected to be considerably less hard than the bare +6 charged ions from which they are theoretically derived. Borderline acids include ions of iron(II), cobalt(II), nickel(II), copper(II), zinc(II), tin(II), lead(II) and antimony(III). Soft acids include ions of copper(I), mercury(I and II), silver(I), gold(I), thallium(I), cadmium(II) and platinum(II and IV). The oxide ion(-2) is considered to be a hard base, sulfide(-2) and arsenide(-3) ions being soft bases.

In general hard acids are ions of metals with low electronegativity coefficient values and which undergo further ionization with difficulty. They have relatively small cationic radii and tend to have high oxidation states. Soft acids are ions of the more electronegative metals and tend to have relatively large cationic radii and low oxidation states. There are only two hard bases, fluoride and oxide ions, which are derived from the two most electronegative elements. Soft bases are associated with the less electronegative non-metals and have larger anionic radii than hard bases. The general tendencies are that hard acids tend to prefer to bond to hard bases and soft acids to soft bases. The hard acid-hard base interaction is predominantly of an ionic nature, whereas the soft acid-soft base interaction is more of a covalent nature.

2.4.2.3 Summary of the discussion of chemical classifications

The lithophiles are those elements which may be described as exhibiting *a*-type character in the (ACD) classification or are hard acids according to the (HSAB) concept. The siderophiles are those elements exhibiting *b*-type character or are soft acids, the chalcophiles being borderline elements in both classifications.

			Co	Ni	Cu	Zn		
Mo				Ag	Cd	In		Sb
					Hg	Tl	Pb	Bi

(a) metal chalcophiles

	Mn	Fe	Co	Ni	Cu	Zn		
Mo		Ru	Rh	Pd	Ag	Cd	In	Sn	Sb
	Re	Os	Ir	Pt	Au	Hg	Tl	Pb	Bi

(b) metals with insoluble sulfides

Fig. 2.4 A comparison of (a) metal chalcophiles with (b) metals having insoluble sulfides; the elements in the enclosure are the siderophiles

2.4.2.4 Sulfide solubilities

A more significant chemical observation is that the chalcophile elements are those whose sulfides are well characterized as being insoluble in water (Burns *et al.* 1980). This is consistent with the supposed hydrothermal production of the chalcophile sulfides. The siderophile elements also have insoluble sulfides (not well characterized) but, because of their general resistance to oxidation, probably lacked the opportunity to exist in such a combined state. Figure 2.4 is a comparison of the metal chalcophiles and those elements with insoluble sulfides. The metals are arranged in the same positions as they have in the periodic classification of the elements. Apart from the siderophile elements the only differences between the (a) and (b) categories of Fig. 2.4 are the appearances of the elements manganese, iron and tin in the latter. The sulfides of manganese(II) and iron(II) possess the highest solubility products of the sulfides of the elements in the (b) category, the lithophilic nature of manganese and iron being in no doubt.

Manganese(II) and iron(II) sulfides, together with those of cobalt(II), nickel(II) and zinc(II), are soluble in acid solutions. The presence of tin in category (b) is consistent with the doubts as to whether or not the element should be classified as a chalcophile or as a lithophile.

A recent book (Williams, 1990) consists of a very extensively detailed and up-to-date discussion of the general reactions of minerals and mineral formation reactions in oxide zones of deposits.

2.5 REFERENCES

Ahrland, S., Chatt, J., & Davies, N.R. (1958) *Quart. Rev. Chem. Soc.*, **12**, 265.
Barrett, J. (1991) *Understanding Inorganic Chemistry*. Ellis Horwood, p.34.
Boldt, J.R. Jnr. (1966) *The Winning of Nickel*. Longmans, (a) p.9, (b) p.191-383.
Brierley, C.L. (1974) *Solution Mining Symposium, 1974*, Aplan, F.F., McKinney, W.A. & Pernichele, A.D. (eds) p.461.
Bryner, L.C. & Anderson, R. (1957) *Ind. Eng. Chem.*, **49**, 721.
Burdett, J.K. & Miller, G.J. (1987) *J. Amer. Chem. Soc.*, **109**, 4081.
Burns, D.T., Townshend, A. & Catchpole, A.G. (1980) *Inorganic reaction chemistry: systematic chemical separation*. Ellis Horwood.
Concha, A., Oyarzun, R., Lunar, R. & Sierra, J. (1991) *Mining Magazine*, p.324.
Derry, R., Garrett, K.H., Le Roux, N.W. & Smith, S.E. (1977) *Geology, Mining and Extractive Processing of Uranium*. Jones, M.J. (ed.) Inst. Min. Metall., p.56.
Duncan, D.W. & Bruynesteyn, A. (1971) *Can. Inst. Min. Trans.*, **74**, 116.
Dutrizac, R.J. & MacDonald, R.J.C. (1974) *Can. Min. Metall. Bull.*, **67**, 169.
Ebner, H.G. & Schwartz, W. (1973) *Erzmetall.*, **26**, 484.
Fleischer, M. (1987) *Glossary of Mineral Species*. 5th edn, Mineralogical Record.
Foo, K.A., Reid, W.W. & Young, J.L. (1990) *Randol Gold Forum, Squaw Valley, 1990*, p.107.
Goldschmidt, V.M. (1937) *J. Chem. Soc.*, p.55.
Groudev, S.N. (1985) *Biogeotechnology of Metals, International Seminar on Modern Aspects of Microbiological Hydrometallurgy, Moscow, 1985*, p.83.
Hoffman, L.E. & Hendrix, J.L. (1976) *Biotech. Bioeng.*, **18**, 1161.
Hunter, J. (1991) *Mining Magazine*, p.58.
Ichikuni, M. (1960) *Bull. Chem. Soc. Japan*, **33**, 1052.
McCready, R.G.L. (1988) *Biohydrometallurgy - 87, Warwick, 1987, Science and Technology Letters*, Norris, P.R. & Kelly, D.P. (eds) p.177.

McCready, R.G.L. & Gould, W.D. (1989) *Biohydrometallurgy - 89, Jackson Hole, 1989*, CANMET SP89-10, Salley, J., McReady, R.G.L. & Wichlacz, P.L. (eds) p.477.

Pearson, R.G. (1963) *J. Amer. Chem. Soc.*, **85**, 3533.

Spencer, P.A., Budden, J.R. & Barrett, J. (1991) *Trans. Instn Min. Metall. (Sect C: Mineral Process. Extr. Metall.)*, **100**, C21.

The Open University (1976), *Mineral Deposits* S266 Block 3, The Open University Press p.20.

Torma, A.E. (1971) *Rev. Can. Biol.*, **30**, 209.

Torma, A.E. & Gabra, G.G. (1977) *Antonie van Leuwenhoek J. Microbiol. Serol.*, **43**, 1.

Trevors, J.T. (1987) *Enzyme Microb. Technol.*, **9**, 331.

Tuovinen, O.H., Hiltunen, P. & Vuorinen, A. (1983) *Eur. J. Appl. Microbiol. Biotechnol.*, **17**, 327.

Tuovinen, O.H., Niemela, S.L. & Gyllenberg, H.G. (1971) *Antonie van Leuwenhoek J. Microbiol. Serol.*, **37**, 489.

Wells, A.F. (1975) *Structural Inorganic Chemistry*. 4th. edn, Oxford University Press. p.605-635.

Williams, P.A. (1990) *Oxide Zone Geochemistry*. Ellis Horwood.

3

The catalytic bacteria

3.1 INTRODUCTION

The three general types of microbiological processes which are of importance to the subject of biohydrometallurgy are:
(i) the oxidation/solubilization reactions which form the classes I, II and III, defined in section 2.1.1, and which are concerned with the solubilization and/or oxidation of sulfide/arsenosulfide minerals and some oxide/carbonate minerals,
(ii) the production of organic compounds by organotrophic microorganisms that are able to solubilize minerals either by the oxidation or the reduction of those elements which have more than one stable oxidation state in aqueous solution, and
(iii) the removal of metal ions from polluted waters by the adsorption of the dissolved species on the surfaces of bacteria (a process known as bioaccumulation) or by precipitation as a result of bacterial action (e.g. the reduction of sulfate ion to sulfide followed by the production of insoluble metal sulfides). Only the first type of processes are considered in this book.

The microorganisms which are important in biohydrometallurgical processes concerned with metal extraction may be divided into the four groups:

(i) mesophiles of the genera *Thiobacillus* and *Leptospirillum*,

(ii) moderate thermophiles of the genus *Sulfobacillus* together with a number of unidentified strains,

(iii) extreme thermophiles belonging to the genera *Sulfolobus*, *Acidanus*, *Metallosphaera* and *Sulfurococcus*, and

(iv) heterotrophic microorganisms.

This chapter consists of descriptions of a selection of representative bacteria which are of relevance to that part of the subject of biohydrometallurgy which is concerned with the bacterial oxidation of minerals. The detailed descriptions are preceded by a general introductory section concerned with the microbiology of bacterial cells, and which defines some of the terminology of the subject.

3.2 GENERAL MICROBIOLOGY

Living organisms are classified as animals, plants or protista. The last group consists of primary or archaic life forms. The protista include those organisms which differ from animals and plants by their lack of form, most of them being unicellular. They are further subdivided on the basis of their cellular structure. The higher protista have cells which are similar to those of animals and plants and are called eucaryotes. Eucaryotes include organisms such as green algae, yeasts and other fungi. The lower protista include bacteria and cyanobacteria (blue-green algae) which form the class of organisms known as procaryotes.

The cell is the smallest unit of which living organisms are composed, and which is capable of life. The basic components of all cells include deoxyribonucleic acid (DNA), ribonucleic acid (RNA), proteins, lipids, phospholipids and polysaccharides. There are fundamental differences between procaryotic and eucaryotic cells. Eucaryotic cells possess a true nucleus which contains the main part of the genome (the DNA which determines the genetic constitution of the cell and its progeny). Minor parts of the genome are contained in organelles such as mitochondria and chloroplasts, the latter being found in plant cells which participate in photosynthesis. Procaryotic cells do not possess a true nucleus. Their DNA is in the form of a single molecule in the cytoplasm (the interior of the cell). It is effectively a single chromosome.

The morphology (form and structure) of the procaryotes is not very varied, bacteria possessing shapes which are either spheres (cocci), straight rods (bacilli) or curved rods (spirilli), with a small range of sizes. This lack of differentiation by shape and size is contrasted by an incredible diversity and adaptability of metabolic characteristics.

Protein synthesis takes place in the cell in the ribosomes. The sequence of bases in the ribosomal RNA of cells has been used to study the phylogeny (evolution) of bacteria. This has led to the theory that primordial cells called progenotes gave rise to two lines of cells, one representing the eubacteria (including Gram-positive and Gram-negative bacteria and the cyanobacteria), the other separating into the eucaryotes and the archaebacteria. The theoretical hierarchy of cell evolution is shown in Fig. 3.1.

The bacteria classed as either Gram-positive or Gram-negative respond differently to the Gram staining reaction. This treatment involves the addition of a solution of the dye, crystal violet, to the cells which have been fixed to a microscope slide. They are then treated with a solution of iodine. Iodine and crystal violet form a complex which does not dissolve in water and is only moderately soluble in ethanol. The cells are then treated with ethanol to differentiate the Gram-positive ones which retain the complex and have a deep blue-purple colour from the Gram-negative ones which are de-stained and have a much paler colour.

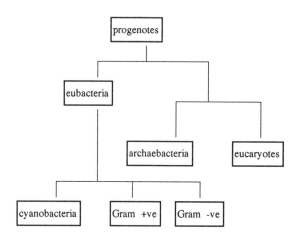

Fig 3.1 Hierarchy of cell evolution

The main components of the procaryotic cell are (i) the cell wall, (ii) the cytoplasmic membrane and (iii) the cytoplasm. These are indicated, together with the periplasmic space, in the schematic diagram in Fig. 3.2.

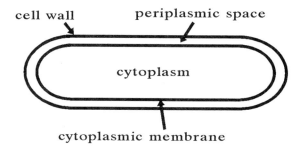

Fig. 3.2 Diagram of the main components of a bacterial cell

Flagella, which are helical protein threads (12-18 nm diameter, up to 20 μm long), are attached to the cell wall of some bacteria and are their means of achieving motility or movement through the medium. Pili, which are long thin threads of protein (3-25 nm diameter, up to 12 μm long) occur in some species and are possibly concerned with the mechanism of attachment of the bacteria to solid surfaces. The main features of the cell are described briefly in the following sections of this chapter and are exemplified in section 3.3 by the details of species which are relevant to the part of the subject of biohydrometallurgy that is concerned with extractive metallurgy.

3.2.1 The cell wall

The cell walls of Gram-positive and Gram-negative bacteria consist of a skeletal material composed of a polymeric substance, murein, which is a peptidoglycan. Short peptides are attached to the peptidoglycan chains and contribute to an intermolecular network. In Gram-positive bacteria the murein network represents 30-70% of the dry weight of the cell wall and consists of up to 40 layers. In Gram-negative bacteria the murein network comprises less than 10% of the cell wall and is present as a single layer. Between the cell wall and the cytoplasmic membrane is the periplasmic space.

The cell walls of archaebacteria do not have a peptidoglycan skeleton. They consist of a 'pseudo-murein' constructed from proteins and polysaccharides and there is no periplasmic space. The relative crudity of their cell walls causes archaebacteria to be somewhat fragile and there is some evidence that they do not survive well in stirred reactors in the presence of fine mineral particles. The attrition posed by the hard mineral tends to destroy the bacterial surface. This disadvantage is not shared by the other bacterial genera used in biohydrometallurgy.

3.2.2 The cytoplasmic membrane

The cytoplasmic membrane provides the effective barrier between the cytoplasm, which contains the metabolic and genetic capabilities of the organism, and the exterior environment. It is a complex structure involving lipids, phospholipids and proteins. It is an osmotic barrier and controls the transport of substances into, and out of, the cytoplasm. The structure of the membrane consists of a bilayer of lipids which is bridged by protein molecules that form pores through which ions and small molecules are transported in a regulated manner. The membrane also contains the enzymes (including cytochromes and iron-sulfur proteins) which are involved in electron transport and oxidative phosphorylation processes that are vital to the metabolism of the cell.

3.2.3 The cytoplasm

The cytoplasm may be regarded as consisting of everything within the cytoplasmic membrane. It may be regarded as consisting of an aqueous medium, whose pH value is about seven, which contains particulate structures. These include ribosomes (sites of protein synthesis), soluble enzymes, RNA, DNA, glycogen and polyphosphate granules, lipid and sulfur inclusions, and possibly plasmids (circular pieces of DNA which are not essential for binary fission, but which contain genes coding for resistance mechanisms, for example to toxic metals). The granules and inclusions are forms of nutrient storage.

3.2.4 Capsules and slimes

It is common for many bacteria to accumulate materials on their outer surface which are referred to as capsules or slimy envelopes. These materials are composed mainly of polysaccharides, although with some bacteria they are composed of polypeptides. The capsular material may be excreted into the medium as slime. Such material is thought to be important in the process of adhesion between a cell and a solid surface.

3.2.5 Systematization of bacterial species

Bacterial species are assigned to groups, the assignments being dependent upon the shape (rods, spirilli or cocci), response to the Gram staining reaction and whether the cells are aerobic or anaerobic. Most bacterial species are assigned to one of twenty such groups, with others awaiting assignment to these groups or to new otherwise unspecified groups.

3.2.6 Metabolism of chemolithoautotrophs

The bacteria of importance in biohydrometallurgy are mainly aerobic chemolithotrophic species. They catalyse the oxidation of some or all of the following species: aqueous iron(II), elemental sulfur, some ions containing sulfur in an oxidation state less than six (e.g. thiosulfate ion, $S_2O_3^{2-}$), and minerals containing iron(II) and/or sulfide. The oxidant (or terminal electron acceptor in microbiological terminology) is normally molecular oxygen which penetrates the cell membrane. Oxygen is reduced to water, the overall reduction process being represented by the half-reaction:

$$O_2 + 4H^+ + 4e^- \longrightarrow 2H_2O \qquad (3.1)$$

Although elemental sulfur has been observed in the cytoplasm of some species, the other substrates (aqueous iron(II) and minerals) are relieved of their electrons in the region of the cell wall, possibly in the periplasmic space in the case of aqueous iron(II). The electrons from the substrate are conducted, by a series of electron transfer catalysts (including cytochromes and iron-sulfur proteins) in the respiratory chain, to appropriate sites in the cytoplasmic membrane where oxygen is reduced as in equation (3.1). As is indicated by the equation the cell has a proton requirement. The overall features of cell metabolism are shown schematically in Fig. 3.3 for the oxidation of pyrite by oxygen. The curved arrows indicate the overall chemical changes occurring to the oxidizing agent (oxygen) and the reducing agent (pyrite), the straight arrow indicating the direction of electron flow. Oxygen and protons can enter the cytoplasm, as indicated by the dotted arrows.

The complex biochemistry which occurs within the cytoplasm, concerned with cell maintenance, growth and division, is not within the scope of this book. It is well documented and is generally regarded as being common to all cells. Further introduction to the subject of general microbiology is to be found in the excellent textbook by Schlegel (1986).

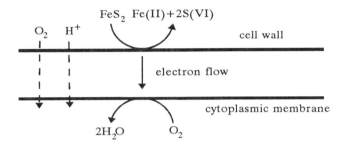

Fig. 3.3 A diagrammatic representation of the bacterially catalysed oxidation of pyrite

3.3 BACTERIAL CHARACTERISTICS

This section consists of a description of the morphology and ultrastructure (the microscopic details of the components of the cell), cell walls, cytoplasm, physiology, and ecology of the more important bacteria used in biohydrometallurgical processes.

Table 3.1 Some properties of *T. ferrooxidans* and *T. thiooxidans*

property	*T. ferrooxidans*	*T. thiooxidans*
cell size/(μm)	0.3-0.5 x 1-1.7	0.5 x 1-2
energy sources	Fe(II), S(-2), S(0), thiosulfate, sulfide minerals	S(-2), S(0), thiosulfate, some sulfides
pH range	1.0-6.0	0.5-6.0
optimum pH	2-2.5	2-2.5
temp. range/°C	2-40	2-40
optimum temp.	28-35	28-30

3.3.1 Genus *Thiobacillus*

Sulfur-oxidizing bacteria were discovered in 1902 by Beijerinck (1904). Since then a large number of species of these bacteria has been discovered. These species are important in the sulfur and iron cycles in the biosphere and can participate in the leaching of metals from minerals. There are at least fourteen species which belong to the genus *Thiobacillus* of which the more important are *Thiobacillus ferrooxidans* and *Thiobacillus thiooxidans*. Some properties of these two species are shown in Table 3.1.

3.3.1.1 Morphology of Thiobacilli

Thiobacilli are chemolithotrophic Gram-negative rod-shaped cells, ranging in diameter from 0.3 to 0.8 micrometres (µm) and in length from 0.9 to two micrometres. A electron micrograph of a *Thiobacillus ferrooxidans* cell is shown in Fig. 3.4.

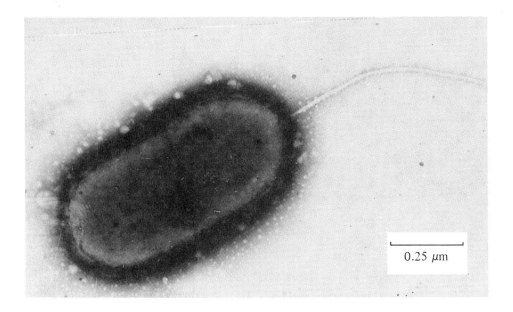

Fig. 3.4 An electron micrograph of a *T. ferrooxidans* cell;
the bar represents a length of 0.25 micrometres

The cell achieves motility by means of a single polar flagellum (shown in the figure). A surface slime layer has been detected in a number of bacteria, and is shown in the example in Fig. 3.4. Pili have been observed on the surface of some *Thiobacilli*. It is thought that they aid the attachment of bacteria to solid substrates.

3.3.1.2 Cell wall

The cell walls of thiobacillus bacteria are typical of Gram-negative species. An electron micrograph of a section through a *Thiobacillus thiooxidans* cell is shown in Fig. 3.5, which demonstrates the complexity of bacterial cells.

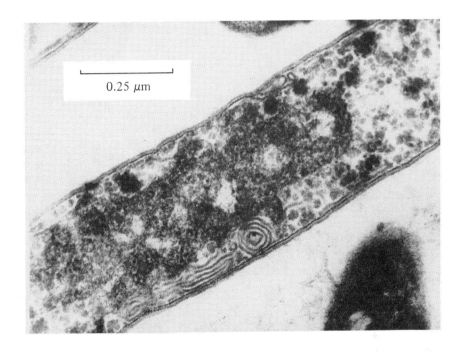

Fig. 3.5 An electron micrograph of a section through a *T. thiooxidans* cell showing some details of its ultrastructure; the bar represents a length of 0.25 micrometres

It includes an external membrane which is 7.5-8.0 nm thick and a peptidoglycan layer from 3-5 nm thick. During the oxidation of elemental sulfur, thiosulfate or tetrathionate ions, granules of sulfur have been observed in the periplasmic space of *T. thiooxidans* and other *Thiobacillus* species. It is possible that elemental sulfur is an intermediate product in the bacterial oxidation of sulfide minerals.

3.3.1.3 Cytoplasmic membrane

The cytoplasmic membrane is a lipid bilayer about 7.5 nm thick and consists of two electron-dense layers about 2.5-3.0 nm thick sandwiching an electron-transparent layer of the same size.

3.3.1.4 Cytoplasm

Various structures occupy an internal part of the cells of thiobacilli. These include ribosomes with sizes from 7 to 19 nm, carboxysomes, and granules of polyphosphates and polysaccharides. Carboxysomes contain ribulose-bisphosphate-carboxylase which is the key enzyme in the fixation of carbon dioxide. Granules of polyphosphates and/or polysaccharides may be found in the cell. The former serve as the cell's reserve of phosphate.

3.3.1.5 Physiology

Thiobacilli are normally strict aerobes and are either obligate or facultative chemolithotrophs or are mixotrophs. *T. ferrooxidans* and *T. thiooxidans* are normally regarded as being obligate chemolithotrophs. Facultative autotrophs (e.g. *T. intermedius*) are able to grow under both autotrophic and heterotrophic conditions. Mixotrophs (e.g. *T. rubellus*) need both inorganic reduced sulfur compounds and organic compounds for their growth. As a source of energy, obligate autotrophs utilize inorganic substrates such as sulfur(0), thiosulfate ion, iron(II), dihydrogen or sulfide minerals. Their sole source of carbon is carbon dioxide.

According to MacIntosh (1976) and Stevens *et al.* (1986), *T. ferrooxidans* is able to fix atmospheric nitrogen to satisfy its nitrogen requirement. This property might explain the activity of *T. ferrooxidans* in the absence of any other obvious nitrogen source.

Bacteria of the genera *Thiobacillus* grow in media of pH values between 0.5 and 10. Some are acidophiles (*T. ferrooxidans* and *T. thiooxidans* in the range 0.5 to 6), others grow at intermediate pH values (e.g. *T. intermedius* in the range 1.9 to 7) whilst some others grow at higher pH values (e.g. *T. thioparus* in the range three to ten). *Thiobacilli* are mesophiles having optimum temperatures for growth at around 30°C. However, they can grow and oxidize inorganic substrates within a wide temperature range between 2°C to 37°C. A peculiar (and essential) property of some acidophilic thiobacilli is their general resistance to comparatively high concentrations of metal ions.

An additional mechanism of sulfur oxidation catalysed by *T. ferrooxidans* (and by *T. thiooxidans* and *Sulfolobus acidocaldarius*) was proposed by Brock and Gustafson (1976) who demonstrated that iron(III) could act as a terminal electron acceptor (oxidizing agent) instead of oxygen. Sugio *et al.* (1985) came to the same conclusion. Pronk *et al.* (1991) showed that iron(III) is the oxidant in the anaerobic *T. ferrooxidans* catalysed oxidation of formic acid. Under aerobic conditions both oxygen and iron(III) were active as electron acceptors. Pronk *et al.* (1991) also showed that formic acid was the only source of energy for growth under otherwise autotrophic conditions. This would imply that *T. ferrooxidans* (or, at least, the particular strain used) should be classified as a facultative autotroph. Fig. 3.6 is a schematic diagram which indicates the possible role of iron(III) as an oxidant during the oxidation of pyrite. For the iron(III) to be the sole oxidant, as it would be under true anaerobic conditions, the oxygen reduction shown in the diagram would be omitted. The electron flow in the anaerobic case would be along the cell wall rather than across it. Although more work needs to be carried out, it may be that the involvement of iron(III) as an oxidant in the primary process is of general applicability.

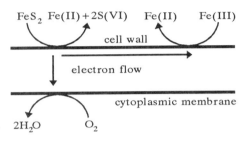

Fig. 3.6 Diagram showing the possible function of iron(III) as an oxidant in the aerobic or anaerobic oxidation of pyrite

3.3.1.6 Ecology

Thiobacilli are widespread in ore deposits, sulfur springs and soils. The identity of the bacterial species may vary with the local temperature and the pH value of the contact water. Bacteria such as *T. thioparus*, as well as heterotrophic bacteria, predominate in waters with neutral and mildly alkaline reaction of the medium. Various kinds of heterotrophs and thiobacilli, including acidophilic examples, are found in great quantities in ores, under conditions of lower pH, while large numbers of acidophilic bacteria are only found in ores and waters with a very low pH value. Distribution of these bacteria also depends upon the type of ore. Favourable conditions for the development of both *T. ferrooxidans* and those acidophiles that oxidize only compounds of reduced sulfur and are resistant to the ions of heavy metals, are created in ores containing pyrite. Ores which have low grades of sulfide minerals contain low population densities of *T. ferrooxidans* ($10^2 - 10^4$ cells per gram rather than the normal 10^{10} cells per gram). While there are many species of the genus *Thiobacillus* and while they all may participate in biogeochemical processes in ore deposits, only *T. ferrooxidans* and *T. thiooxidans* have any practical importance for biohydrometallurgy.

Several species of moderately thermophilic bacteria have been discovered in recent years. They are similar to thiobacilli but, since their position in the systematic description of bacteria has not been fully studied, they are dealt with in a separate section (3.3.3).

3.3.1.7 Associates of T. ferrooxidans

The presence of acidophilic, heterotrophic, bacteria in hitherto presumed pure cultures of *T. ferrooxidans* has received much attention. Zavarsin (1972) detected a heterotrophic associate in a *T. ferrooxidans* culture which consisted of thin motile rods. The associate grew at pH values up to 2-3 in the presence of iron(II) when citric acid was added to the medium. Less active growth was observed in the medium in the presence of sucrose, fructose, ribose, glucose, maltose, xylose, fumaric and succinic acids, mannitol and ethanol. Its taxonomical position is not clear. According to Zavarsin it is similar to *Acetobacter acidophilum*. Guay and Silver (1975) isolated cultures of a new acidophilic bacterium *T. acidophilus* from cultures of *T. ferrooxidans*. This bacterium was capable of autotrophic growth on elemental sulfur in addition to heterotrophic growth on a variety of organic compounds. Harrison *et al.* (1980) showed that some stock cultures of *T. ferrooxidans* contain a heterotrophic associate which has been isolated and named *Acidiphilium cryptum*. It is a Gram-negative aerobic, mesophilic rod-shaped bacterium that grows on weak organic media in the pH range 1.9-5.9.

Such associates may be quite common and could be of significance to the efficient functioning of autotrophs under autotrophic conditions by the heterotrophs metabolizing the organic products of the autotrophs.

3.3.2 Genus *Leptospirillum*

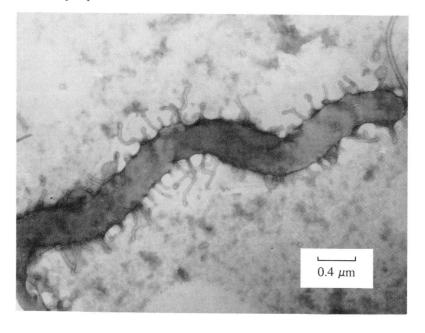

Fig. 3.7 An electron micrograph of a *L. ferrooxidans* cell; the bar represents a length of 0.4 micrometres

The genus *Leptospirillum* is part of the family of *Spirillaceae* and includes one mesophilic species, *L. ferrooxidans*, which was isolated from an Armenian deposit (Markosjan, 1972). More recently a number of new mesophilic strains of bacteria similar to *L. ferrooxidans* have been isolated from solutions used in metal leaching (Harrison and Norris, 1985, Johnson *et al.* 1989).

A moderately thermophilic strain was isolated (Golovacheva *et al.* 1992) and has been classified as the new species *Leptospirillum thermoferrooxidans*. Whereas the optimum temperature range for *L. ferrooxidans* is around 30°C, that of *L. thermoferrooxidans* is between 45-50°C and their optimum pH ranges are 2.5-3 and 1.65-1.9 respectively.

All leptospirilli are strict aerobes and gain energy solely by oxidizing iron(II) in aqueous solution or the iron(II) content of minerals. An electron micrograph of a *Leptospirillum ferrooxidans* cell is shown in Fig. 3.7. The polar flagellum is featured, together with slime production along the sides of the cell.

Fig. 3.8 is an electron micrograph of cells of *Leptospirillum thermoferrooxidans* with their polar flagella and slime formation indicated.

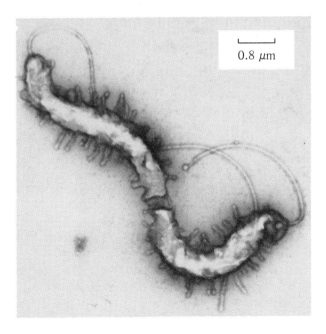

Fig. 3.8 An electron micrograph of *L. thermoferrooxidans* cells showing their flagella and slime; the bar represents a length of 0.8 micrometres

The cells, by accumulating iron precipitates, can assume a coccoid-like appearance as shown by the electron micrograph in Fig. 3.9. The cell wall structures of *L. ferrooxidans* and *L. thermoferrooxidans* are similar to those of other Gram-negative bacteria. The cells accumulate materials on their outer surface which leads to the formation of capsules (the bacterial cells are encapsulated by the material) or slimy envelopes. The capsular material may be excreted into the medium as slime. The formation of capsules could account for the accumulation of iron compounds on their surface which leads to the formation of a coccoid-like appearance. Both species are obligate autotrophs and derive energy only from the oxidation of iron(II).

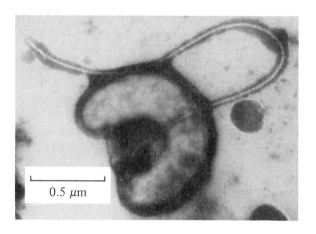

Fig. 3.9 An electron micrograph of a *L. thermoferrooxidans* cell which has accumulated an iron-containing precipitate; the bar represents a length of 0.5 micrometres

3.3.2.1 Other bacteria of the genus Leptospirillum

Other *Leptospirillum*-like bacteria were studied by Harrison and Norris (1985). Strains similar to those of *L. ferrooxidans* were isolated from different habitats and known by their code names: BU-1 from a copper mine in Bulgaria, CH from leached dumps of copper ore, BC and ALV from coal dumps and LAM from a uranium deposit in Mexico. They are curved rods with one polar flagellum. They form aggregates of cells in slimes.

These bacteria are widespread in ore deposits, in ore dumps and in the solutions produced in the process of copper leaching. They also always accompany *T. ferrooxidans* in the reactor leaching of sulfide concentrates. The thermophilic strains are constant companions of bacteria belonging to the genus *Sulfobacillus* (described in section 3.3.3.1).

3.3.3 Moderately thermophilic bacteria

The existence of moderately thermophilic acidophilic bacteria in thermal springs and ore deposits was demonstrated by Le Roux *et al.* (1977) (code name: TH1), Brierley, J.A., and Lockwood, (1977) (code name: TH2) in the United States and by

Golovacheva and Karavaiko (1977) in Russia. Other cultures in this class have been discovered, the most recently published example being isolated in Australia in 1986 by Poole (code name: M4) (Nobar *et al.* 1988). Some of the species have been classified as being members of the new genus *Sulfobacillus*, others are still known only by their code names.

3.3.3.1 Genus Sulfobacillus

The bacterium isolated from ore dumps by Golovacheva and Karavaiko in 1976 was first described as a thiobacillus-like organism. A more detailed study led to its being designated as a member of a new genus named *Sulfobacillus*. This genus is composed of aerobic Gram-positive facultatively autotrophic eubacteria which are able to use iron(II), elemental sulfur and reduced sulfur compounds as energy sources. The initially discovered species was named *Sulfobacillus thermosulfidooxidans* and other sub-species have now been described; e.g. *S. thermosulfidooxidans* subspecies *thermotolerans* (Kovalenko and Malakhova, 1983) and *S. thermosulfidooxidans* subspecies *asporogenes* (Karavaiko *et al.* 1988).

The bacteria of the genus are physiologically similar. Their sources of energy are iron(II), sulfur and minerals including pyrite, chalcopyrite, arsenopyrite, sphalerite, antimonite, covellite and chalcocite. Autotrophic growth is supported by thiosulfate ion. Better growth is observed if the medium contains 0.01-0.02% yeast extract. All the species grow well under mixotrophic conditions with various minerals in the presence of yeast extract, some sugars, amino acids or more complex organic substrates such as glutathione or hydrolyzate of casein. They also grow under heterotrophic conditions on the same substrates (i.e. the organic substances in the absence of the inorganic minerals). The serial sub-culturing on the organic substrates does not cause the loss of the ability to catalyse mineral oxidation. These observations lead to the characterization of the species as facultative autotrophs. They are all strict aerobes and extreme acidophiles. Their optimum temperatures for growth lie between 37-42°C for *S. thermosulfidooxidans* subsp. *thermotolerans* to 50°C for the other two species, with maximum temperatures which allow growth of 58-60°C for all three species.

Bacteria of this genus are widespread in nature. They occur in dumps of sulfide ores and in volcanic regions. *S. thermosulfidooxidans* was found in the corrosion zones of pipelines in a city heating system, and in 70% of ore samples from dumps subjected to leaching. The cell numbers in warm run-off solution from dumps amounted to 2.5×10^6 cells mL^{-1} which is an indication of the participation of the microorganisms in processes of mineral oxidation and metal leaching.

The three members of the genus are morphologically similar in that they consist of straight rods although other forms are sometimes observed. Fig. 3.10 is an electron micrograph of a cell of *Sulfobacillus thermosulfidooxidans* subspecies *thermotolerans*.

3.3.3.2 Non-classified moderate thermophiles

Brierley and Le Roux (1977) isolated an aerobic bacterium from hot springs in Iceland which oxidizes iron(II), pyrite, pentlandite and chalcopyrite. Active growth of the organisms took place only in the presence of yeast extract. Growth occurred at temperatures between 30-50°C and ceased at 60°C. A number of strains were isolated from thermal springs in Iceland and from dumps during copper leaching. They were given the code names TH1 (Le Roux *et al.* 1977), TH2 (Brierley,

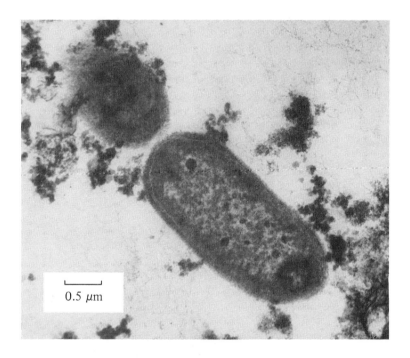

Fig. 3.10 An electron micrograph of a *Sulfobacillus thermosulfidooxidans* cell; the bar represents a length of 0.5 micrometres

J.A. 1978) and TH3 (Norris *et al.* 1980). The bacteria are rods with sizes 0.8 μm x 1.6-3.2 μm, 0.8 μm x 1.6-4.2 μm and 0.5 μm x 1-1.6 μm respectively. They all grow at an optimum temperature in the region of 50°C and in an iron(II) medium with a pH range between 1.4 to 3. They do not catalyse the oxidation of elemental sulfur. Either organic substances or carbon dioxide serve as carbon sources and the bacteria require either yeast extract or organic substances containing sulfur (e.g. glutathione or cysteine) to provide energy. On yeast extract they grow under heterotrophic conditions.

A number of strains of thermophilic aerobic iron and sulfur oxidizing bacteria (code name: LM) have been isolated from samples of sediments and water from Lake Myvam in Iceland (Marsh and Norris, 1983). Some of them oxidize sulfur at 50°C and are similar to *T. thiooxidans*. Another isolate, growing on a medium containing iron(II) at 50°C, is similar to *T. ferrooxidans*. Optimum values of pH for growth were between 1.5-1.8.

Hendy (1987) isolated a thermophilic rod-like bacterium from the ores of the Rum Jungle deposit in Australia. The culture oxidized iron(II) and elemental sulfur at between 50-55°C, the maximum temperature used being 60°C. The strain grows with the addition of yeast extract at an optimum pH value of 1.4. Very similar strains were isolated from a coal spoil tip at Alvcote (code name: ALV), a washed coal pile at Birch Coppice (code name: BC) (Marsh and Norris, 1983) and from a spoil heap at Kingsburg (code name: K) (Wood and Kelly, 1983). These bacteria grow on a mineral medium containing iron(II) and utilize carbon dioxide as their carbon source even in the presence of glucose. They grow on organic substrates only in the presence of iron(II). The ALV strain can assimilate sulfate ion and the LM2 strain oxidizes elemental sulfur.

The mixed thermophilic culture, known by the code name, M4, discovered by Poole in Australia, has been described by Nobar *et al.* (1988). It consists of two differently sized rods and a coccus and was shown to catalyse the oxidation of pyrite, arsenopyrite, chalcopyrite and pyrrhotite with an optimum temperature of 46°C in the pH range 1.3 to 1.6. The rods are Gram-negative, the small ones being iron(II) oxidizers, the large ones being heterotrophic sulfur oxidizers. The cocci are Gram-positive and do not oxidize sulfur. The three components have survived many adaptations during the treatment of around twenty different substrates in various countries and appear to exist as a resilient interdependent symbiotic community.

Like bacteria of the genus *Sulfobacillus* all the above species are characterized by their ability to grow under autotrophic, mixotrophic and heterotrophic conditions and are mainly dependent upon reduced sulfur compounds. It is quite possible that some or even all these moderately thermophilic species will be assigned to the genus *Sulfobacillus* when they have been fully characterized.

3.3.4 Thermoacidophilic archaebacteria

Archaebacteria represent an independent branch in the evolution of microorganisms. These bacteria possess a number of unique properties which differentiate them from eubacteria, of special significance being the absence of peptidoglycan in the cell wall. There are four genera of these bacteria which oxidize compounds of sulfur. They are *Sulfolobus*, *Acidanus*, *Metallosphaera* and *Sulfurococcus*. They are all aerobic, extremely thermophilic and acidophilic bacteria which are coccoid in form. The cells are immotile and have no flagella. They have a negative response to the Gram staining test but should not be described strictly as Gram-negative bacteria. An electron micrograph of a single cell of *Sulfolobus yellowstonii* is shown in Fig. 3.11.

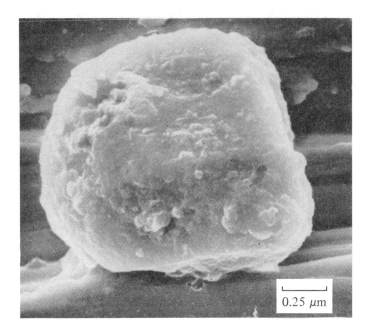

Fig. 3.11 An electron micrograph of a *Sulfolobus yellowstonii* cell; the bar represents a length of 0.25 micrometres

3.3.4.1 Morphology

The only species which are of any importance in biohydrometallurgy are *Sulfolobus* and *Acidanus*. They are coccoid bacteria with diameters of around 1 µm. *Sulfolobus* species have pili-like structures on their surfaces. These are suspected to be involved with the mechanism of the attachment of the cells to surfaces. Their cell walls consist of glycoprotein and do not have a murein layer.

3.3.4.2 Physiology

All the species are facultative chemolithoautotrophs and grow under autotrophic, mixotrophic or heterotrophic conditions. Under autotrophic conditions the bacteria catalyse the oxidation of elemental sulfur, iron(II) or sulfide minerals and gain energy from the process. They use carbon dioxide as their carbon source. They grow more rapidly under mixotrophic conditions in the presence of 0.01-0.02% yeast extract or other organic substances. *S. acidocaldarius* oxidizes elemental sulfur under anaerobic conditions utilizing iron(III) as a terminal electron acceptor (oxidizing agent) (Brock and Gustafson, 1976). Both *A. brierleyi* and *S. acidocaldarius* have the capacity of using molybdenum(VI) as a terminal electron acceptor under similar conditions (Brierley and Brierley, 1982). *S. acidocaldarius* can grow in the pH range 1-5.9 with an optimum range of 2-3, the optimum range for

A. brierleyi being 1.5-2. The temperature ranges for the two species are 55-80°C and 45-75°C respectively, both having an optimum growth rate at 70°C. Some details of the ultrastructure of the cell are demonstrated by the electron micrograph of a section shown in Fig. 3.12.

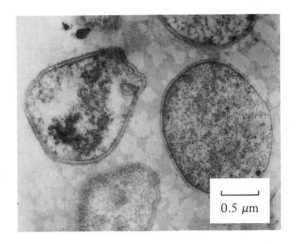

Fig. 3.12 An electron micrograph of sections through cells of *Sulfolobus yellowstonii* showing their ultrastructure; the bar represents a length of 0.5 micrometres

3.3.4.3 Ecology

The species inhabit high temperature sulfur springs. Representatives of the genera *Sulfolobus* and *Acidanus* were isolated from the thermal springs of Yellowstone National Park (Wyoming, USA) in waters at temperatures between 43-99°C and with pH values in the range 1-5.9. Other species have been discovered in the volcanic regions of Iceland, Italy, the Azores, New Zealand, Japan, the Kuril Islands and Kamchatka.

3.3.5 Guanine and cytosine content of bacterial DNA

The four bases in DNA chains are adenine, cytosine, guanine and thymine. One parameter which is used in the classification of bacterial species is the sum of the percentages (in molar terms) of the bases guanine and cytosine in the DNA of their cells compared to the total of the four bases present. Some typical values of the mole percentage of guanine and cytosine for the bacterial species discussed above are shown in Table 3.2.

Table 3.2 The guanine (G) and cytosine (C) content of various genera

species	(G + C) mol %
T. ferrooxidans	55-57
T. thiooxidans	50-53
L. ferrooxidans	50
L. thermoferrooxidans	65
S. thermosulfidooxidans	47-49
moderate thermophiles	49-68
Sulfolobus	34-39
Acidanus	31-39

3.4 CATALYTIC ACTIVITY OF BACTERIA

The main mechanistic considerations relevant to biohydrometallurgical operations are those concerning the oxidations of iron(II) in aqueous solution and of iron(II) and sulfur (in the formal oxidation states -1 as S_2^{2-} and/or -2 as S^{2-}) in minerals. Aqueous iron(II) is the most easily oxidized substrate for *T. ferrooxidans* and a number of acidophilic archaebacteria, and is the sole source of energy for *Leptospirilli*. The standard reduction potential for the iron(III)/iron(II) couple in 1 M acidic solution is 0.77 V under which conditions the ions are unhydrolyzed. The effects of hydrolysis and complex formation are discussed in Chapter 4, the values of the standard reduction potentials being somewhat lower than 0.77 V, but still higher than those observed for minerals in contact with aqueous solutions. It is significant that the bacterial oxidation of aqueous iron(II) is faster than any of the mineral oxidations. The oxidation of aqueous iron(II) by oxygen is thermodynamically feasible at pH values below six but, at pH values above 1.8, the iron(III) product is not water soluble. This has implications for the site of the oxidation process. If the reaction takes place within the periplasmic space the precipitation of iron(III) could occur if the local pH value is greater than 1.8. The oxidation rates of minerals are considerably lower than that of aqueous iron(II) and are significantly dependent upon the nature of the mineral.

The primary process of the bacterial oxidation of a mineral involves the adsorption of the bacteria onto the sulfide mineral surface. There is no question of the mineral entering the cell, or even the periplasmic space, the rate determining step essentially consisting of the electron transfer from the mineral to the electron transport chain of the bacterium. This may be mediated by the presence of iron(III) species either on the mineral surface or on the bacterial surface or both. In the presence of bacteria iron(III) accelerates the oxidation of practically all sulfide minerals.

In order to catalyse the oxidation of sulfidic minerals there must be a mechanism for the reactants and the bacteria to come into close contact. This section is devoted to a consideration of the factors which could be responsible for the association of bacterial cells with their oxidizable substrates, together with possible factors which affect the rates of the oxidation reactions.

3.4.1 Interaction of bacteria with surfaces

The attachment of bacteria to solid surfaces is the most important process connected with their mode of action. It is essential for the action of the bacteria as catalysts, but is not normally rate determining. Many mechanisms of the adhesion of microorganisms to surfaces have been suggested, reflecting the wide variety of microorganisms, substrates and growth conditions. Adhesion characteristics can vary between different strains of the same organism. This phenomenon may be due to the adaptations undergone by these strains. The influence of physical factors on the adhesion of microorganisms to surfaces is important. Figs. 3.13 and 3.14 show electron micrographs of *T. ferrooxidans* and *S. thermosulfidooxidans*, respectively, in contact with sulfide minerals. Rossi (1990a) has summarized investigations of bacterial adhesion to surfaces and concludes that none of the mechanisms allows the full understanding of the interaction of chemolithoautotrophic bacteria with their substrates.

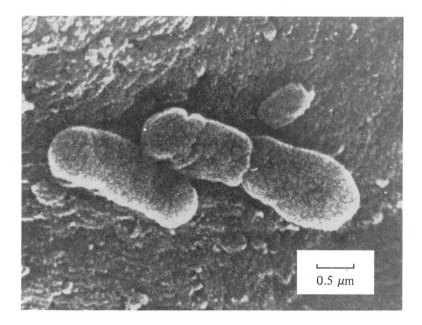

Fig. 3.13 An electron micrograph of cells of *T. ferrooxidans* in contact with a mineral surface; the bar represents a length of 0.5 micrometres

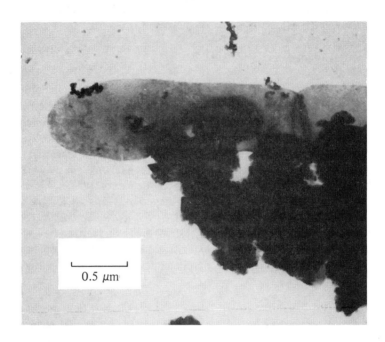

Fig. 3.14 An electron micrograph of cells of *S. thermosulfidooxidans* in contact with mineral crystals; the bar represents a length of 0.5 micrometres

As sulfur and sulfide minerals are sources of energy for chemolithoautotrophs, specific interactions between the microorganisms and the mineral might be expected. It has been known for some time that chemolithoautotrophs become attached to their substrates. It is accepted that oxidation of sulfur by *T. thiooxidans* or sulfur and sulfide minerals by *T. ferrooxidans* proceed by the close contact of the bacteria and their substrates. When measuring cell protein Gormely and Duncan (1974) showed that 65% of the *T. ferrooxidans* population was attached to the particles of zinc sulfide concentrate during its leaching. Dispirito et al. (1981, 1983) showed that between 61-82% of *T. ferrooxidans* cells were adsorbed by various particles (including sulfur and pyrite). In similar experiments with chalcopyrite and sulfur 96-98% and 77% of cells were adsorbed on the respective solids. Wakao et al. (1984) demonstrated that cells of *T. ferrooxidans* were adsorbed very quickly on particles of pyrite. Electron microscopy has shown aggregative adsorption on restricted areas of pyrite particles. Bagdigian and Myerson (1986) showed that *T. ferrooxidans* congregates selectively at dislocations, grain boundaries and other discontinuities on pyrite surfaces. Berry and Murr (1976) reported the electrostatic adhesion of thermophilic bacteria to sulfide minerals. The cell distribution on sulfur and sulfides depends upon the stage of growth of the population and the degree of substrate oxidation. Quantitative measurement of the population of *T. thiooxidans* during its growth on sulfur that at the end of the lag phase about 70% of cells were adsorbed onto the sulfur substrate but only 40% were adsorbed after 168 hours. On the whole, from 60-90% of *T. ferrooxidans* cells are attached to sulfide minerals being treated. There is little doubt that a dynamic equilibrium exists in the pulp between the cells on the solid surface and those in the liquid phase.

Adhesion occurs in stages and depends upon the physiological state of the bacteria. It has been demonstrated that after one hour of *T. ferrooxidans* adsorption to the particles of a copper-zinc concentrate, the cells are easily washed off. However, after several hours the bond between the bacteria and the mineral surface becomes more stable. It could be that the adhesion time of bacteria to their substrates determines the length of the lag phase. The adhesion activity of *T. ferrooxidans* to sulfide minerals depends upon the extent of their adaptation to the substrate. For adapted strains the equilibrium between adsorbed and free cells was established after only 1-5 hours, whereas unadapted cells required 120-300 hours. The attachment of *Sulfolobus* to crystals of sulfur via adhesive pili has been demonstrated (Weiss, 1973). Abundant slime formation is observed in the regions of minerals where there is cell adhesion. Cook (1964) reported that moistening agents promote the contact between bacteria and sulfur. Even after a short time of contact between a culture and a mineral substrate there are noticeable changes in the electrode potential and flotation properties of the mineral. In the latter case the mineral is rendered passive. The analysis of the surfaces of sulfur and sulfide minerals in places of long term adhesion of bacterial cells has revealed the presence of corrosion zones.

The adsorption of *S. acidocaldarius* on pyrite in coals (Vitaya and Toda, 1991) exhibited two stages. The first was a rapid physical adsorption over one minute followed by a second stage over thirty minutes associated with the oxidation of the pyrite surface.

Considering the evidence available it would seem that the interaction between a culture and a mineral surface consists of two stages. The first stage could be described as the physisorption of the bacteria on the mineral surface by an electrostatic interaction. The second stage could be described as chemisorption, where the bonding between the cells and the mineral becomes stronger and more permanent because of the formation of chemical bonds. One example of this would be the formation of sulfur-sulfur bonds between the mineral and cell surfaces.

There are two general factors which influence the interaction between a bacterial cell and a mineral surface. These are (i) the electrostatic charge on the cell surface and (ii) the existence of exometabolites which are substances excreted by the bacteria.

3.4.1.1 Surface charge of bacteria

The surface of a microbial cell is charged due to the presence of groups such as carboxyl (-COOH), amino ($-NH_2$) and hydroxyl (-OH) in the cell wall material. The carboxyl group may lose a proton to become a negative carboxylate, the amino group may be protonated to become positively charged. Protein side chains play the most important role in the formation of the surface charge of cells. The charge on Gram-positive bacteria is mainly determined by the ionization of proteins of the cell walls and any flagella. In the case of Gram-negative bacteria the charge is determined mainly by the components of lipopolysaccharides, the lipoproteins of the cell wall and by the proteins of some enzymes, flagella and pili. The role of the surface charge of cells is crucial to their interaction with their environment. To a great extent the surface charge is determined by the pH of the bacterial medium. At the pH at which most bacterial oxidation reactions occur (0.5-1.5) the nitrogen centres of proteins and the carboxyl groups of proteins and polysaccharides are protonated (as are the hydroxyl groups of polysaccharides) so that the bacterial surface would be expected to be positively charged. It is thought that such a positive surface charge is partly responsible for the cell's ability to resist metal ion toxicity, the metal ions usually being positively charged. It is also possible under some circumstances for there to be a layer of negative ions (largely hydrogen sulfate ions) loosely associated with the positive bacterial surface so that the cell could appear to have a negative charge. For an aqueous ion to enter the periplasmic space the initial stage could be its replacement of protons in the cell wall. The surface charge of the cells is probably the main factor influencing the initial interaction between them and their substrates.

3.4.1.2 Exometabolites

During the oxidation of sulfur and/or metal sulfides bacteria release organic substances into the medium. These substances are known as exometabolites. Their secretion correlates with the bacterial growth and reaches a maximum in the exponential growth phase. Some have high relative molecular mass (RMM) values and include lipids. Amongst the low RMM substances there have been observed organic acids of the tricarboxylic acid cycle, amino-acids (aspartic acid, serine, glycine, alanine and valine) and ethanolamine. The function (if any) of these exometabolites has not been studied extensively. It is possible that lipids and phospholipids could act as moisturizing agents and contribute to the oxidation of sulfur, converting it to a colloidal state so that it could be transported into the periplasmic space. The low RMM substances could participate as complexing agents for ionic species such as aqueous iron(II) and arsenic(III).

It is likely that the production of exometabolites, together with any pili of the bacteria and other surface features, are responsible for the second stage in the adhesion of the cells to their substrate surfaces.

3.4.2 Factors influencing reaction rate

Factors which could possibly influence the rate of oxidation, for any given bacterial species are:

(i) the nature of the mineral,

(ii) the solubility of the mineral,

(iii) the type of semi-conduction (normal or positive hole) exhibited by the mineral,

(iv) the electrode potential between the mineral and the aqueous solution, and

(v) the energy of the Fermi level (effectively the ionization energy) of the mineral.

The influence of solubility and the semi-conduction properties of non-ferrous sulfides upon their bacterial activity has been studied by Tributsch and Bennett (1981a,b). The study includes sulfides which are insulators (ZnS), semi-conductors (CuS, Cu_2S) and sulfides which exhibit metallic conduction (NiS). They include sulfides with cubic structures (MnS, CoS) and with layer structures (WS_2 and MoS_2). Sulfides with n- and p-type conduction were studied and which had different solubility products. *T. ferrooxidans* catalysed the oxidation of all the sulfides. No specific factor emerged from the work which could be directly related to the rate of oxidation of the solid substrate. There is little doubt that the rate is dependent upon the nature of the metal component of the sulfide but no indication of why this should be the case from the factors studied. The vital factor may be the ionization energy of the mineral surface (equivalent to the Fermi level of a metal) which could be determined by X-ray photoelectron spectroscopy.

There have been various attempts to relate the rate of bacterial oxidation of a mineral substrate to the potential developed between the mineral and an aqueous solution. It should be made clear that such attempts are theoretically and experimentally flawed. Electrode potentials are thermodynamic quantities concerned with the initial and final states of a system and normally have no connexion with the rate of progress from one to the other. They determine the position of equilibrium of a system and not the rate at which equilibrium is attained. The rate of a reaction is largely determined by the magnitude of the energy barrier (activation energy) between the initial and final states. There is no doubt that the aqueous solution in a bacterial oxidation reaction has a measurable redox potential. If this potential is greater than that between the mineral and the solution the oxidation of the mineral should proceed, but implies nothing concerning the rate of the process. If the mineral potential is greater than that of the solution the reaction will not proceed. It is not clear what determines the value of the mineral electrode potentials but the nature and state of the mineral surface must be very influential. The experimentally observed values (e.g. Yakhontova, 1985) for mineral potentials vary from the standard values calculated (Barrett, 1992) from thermodynamic quantities as may be seen from Table 3.3. Experimental values for mineral potentials vary considerably and show very little consistency. The values are appreciably dependent upon the state of the mineral surface and the method of measurement. Without a standardized procedure for obtaining these measurements (surface treatment of the mineral and the composition of the aqueous solution) very little can be concluded from their study.

Table 3.3 Calculated (standard) and observed values of reduction potentials, E, of some minerals

mineral	$E^{\ominus}$(calculated)/V	E(observed)/V
sphalerite	0.32	0.35
chalcocite	0.427	0.35
chalcopyrite	0.353	0.4
pyrrhotite	0.27	0.45
arsenopyrite	0.219	0.5
pyrite	0.351	0.55-0.6

3.4.3 Bacterial growth characteristics

Bacterial growth normally takes place by binary fission and is characterized by the doubling time, which is the time taken for the bacterial population to double in the exponential phase of growth. An alternative parameter to the doubling time for a species is the specific growth constant, μ, (equivalent to a first order rate constant for a chemical reaction) which is the natural logarithm of 2 (0.693) divided by the doubling time and has units of reciprocal time (usually h^{-1}). A number of models have been created which attempt to describe bacterial growth and the oxidative processes that the bacteria catalyse. They are fully discussed in the book by Rossi (1990b), the general conclusion being that none of the models is satisfactory. It is not the aim of this book to discuss these models any further but it is of interest to consider the factors which affect the doubling time of the bacteria which are important in biohydrometallurgical processes.

The precision with which doubling times may be estimated is not very high, due to the difficulties of measuring the bacterial population (see Chapter 9) and of the uncertainties associated with the concentrations of soluble substrates and surface areas of solid substrates. A further complication, which may be of major significance, arises from the difficulties associated with arranging conditions under which there is only one substrate present in any system. Bearing these serious reservations in mind there are some generalizations which can be made concerning the effect of the substrate on the doubling time for a given species.

T. ferrooxidans, grown on iron(II), exhibits doubling times between 3.6 and twelve hours. Using a sulfur substrate this rises to between ten hours and eight days, whereas the sulfur oxidizer, *T. thiooxidans*, grows on the element with a doubling time of between ten hours and one day. *T. ferrooxidans* grown on various sulfide minerals exhibits doubling times of up to two days.

The obligate iron oxidizer *L. ferrooxidans* grows on aqueous iron(II) with a doubling time of between nine and thirty-two hours, and the moderate thermophile, *L. thermoferrooxidans* doubles its population in around thirty hours under similar circumstances. The non-identified moderately thermophilic bacteria have doubling times between thirteen and twenty-nine hours when growing upon aqueous iron(II).

The moderate thermophile, *S. thermosulfidooxidans* subsp. *asporogenes*, grown on aqueous iron(II), has a doubling time of 11.7 hours which is reduced to 6.9 hours if the medium contains 1 mM glucose. Its doubling time when growing upon chalcopyrite is 86.6 hours and this is reduced to 43.3 hours if the medium contains 1 mM glucose. The presence of other organic substrates (yeast extract, amino acids and other sugars) have similar enhancing effects upon the growth characteristics. The organic additives also increase the rate of oxidation of the mineral substrate. The effects upon bacterial growth and rate of mineral oxidation are not necessarily directly linked.

The doubling times of *Sulfolobus acidocaldarius* strains are between 6.5 and eight hours if 0.1% of yeast extract is present. The Lake Myvam *Sulfolobus* exhibited doubling times of nine, ten and sixteen hours, when grown on aqueous iron(II), sulfur and pyrite, respectively.

Considering the diversity of the bacterial species, the varying conditions and temperatures, there is relatively little variation between doubling times for different bacteria grown on the same substrate. Also the doubling times for growth upon mineral sulfides are the highest. It is possible to conclude that the most important rate determining factor, both for bacterial growth and mineral oxidation, is the nature of the mineral.

3.4.4 Conclusions

The major conclusion to be drawn from the above discussions is that the rate determining factor, in the bacterially catalysed oxidation of minerals, is the nature of the mineral. The only measurable quantity which could characterize the rate of oxidation for a particular mineral is the ionization energy (Fermi level). There seems to be little doubt that iron(III) is intimately involved in the rate determining step of the primary oxidation process. The adhesion process occurs in two stages; a rapid physisorption stage followed by a slower chemisorption stage, neither being rate determining for the overall oxidation process.

3.5 TOXIC EFFECTS OF METALS

Acidophilic chemolithoautotrophic bacteria are characterized by a high resistance to the toxic effects of metal ions compared to that shown by other bacteria. The principles underlying the resistance of *T. ferrooxidans* and other bacteria to the ions of metals have been formulated by Norris and Kelly (1978). In essence these are that the toxicity of metals depends upon (i) the physiological state of the bacteria, and (ii) the oxidation states and chemical forms (i.e. speciation, see Chapter 4) of the metals which govern their bioavailability. The resistance of wild-type strains of iron(II)- and/or sulfide-oxidizing bacteria to metal ions varies with their habitat and the extent of their adaptation under natural conditions. Strains more resistant to the toxic effects of metal ions may be isolated from environments associated with the leaching of ores or concentrates, since elevated concentrations of metal ions in general are likely to be present in addition to those of toxic metals solubilized from the mineral. Chemolithotrophic bacteria in general adapt fairly readily to such technologically important conditions.

High resistances of bacterial strains to the ions of metals cannot be explained only by physiological or genetic properties of the bacteria. It is rather a pseudoresistance which depends also on the metallic ion or ions in each different environment.

While examples of plasmid-encoded resistance mechanisms are known for chemolithoautotrophs the general tolerance of these bacteria to high metal ion concentrations may also be attributed to chemical factors associated with their growth conditions. These include:

(i) the low pH values of their optimal media, which results in the protonation of anionic sites on cell walls which are then less available for the binding of toxic metal cations,

(ii) the presence of anions which may lead to the precipitation of the metal ion (phosphate and arsenate ions in particular) or its conversion into a less bioavailable complex,

(iii) the presence of organic ligands secreted by cells or released from dead cells, which may complex the metal ion and decrease its bioavailability, and

(iv) the presence of substances which compete with, and thereby alleviate the toxicity of, the toxic metal or metalloid species; for example, phosphate ion decreases the toxicity of arsenate ion, and potassium ion that of thallium(I) ion. High concentrations of the competing species are usually required to achieve the last effect. In general the toxic effects of metal ions are likely to be decreased as the complexity of the medium is increased.

Examples of plasmid-encoded resistance mechanisms include those for Hg^{2+} and UO_2^{2+} in *T. ferrooxidans*. It should be noted that most strains of *T. ferrooxidans* contain plasmids but their function has not yet been characterized. The use of standard genetic manipulation techniques to transfer resistance to *T. ferrooxidans* has been complicated by difficulties in growing up cultures with high cell densities and in plating out the culture onto solid media. These matters are now being resolved but the basic issue of how to return the plasmids containing the genes for metal resistance into the cell is still a major problem. A general discussion of the current state of understanding of the genetics of bacterial metal resistance is given by Mergeay (1991).

However, as implied above, it is usually possible to build up the tolerance of a particular bacterial species by regularly subculturing it at increasingly higher concentrations of the toxic metal species, as this allows the selection of cells which are naturally resistant to these toxic effects. A good example of adaptation of bacteria to copper(II) and nickel(II) ions is given by Natarajan *et al.* (1983) and Natarajan and Iwasaki (1983). They developed separate copper- and nickel-resistant strains and then employed them separately in leaching experiments using bulk copper-nickel concentrates. Their results are given in Table 3.5. The copper and nickel tolerant strains caused large improvements in the leaching of their respective metals. They were still inactive at leaching the other metal. Although there are many reports of the levels of various metals which affect the growth and leaching behaviour of a variety of chemolithotrophs it is difficult to compare data from different sources as no account has been taken of the speciation of the metal ions, which may vary under different conditions.

Table 3.5 Effects of adaptation of bacteria to copper and nickel tolerance; results after forty five days

conditions	% Cu leached	% Ni leached
sterile	1.7	20
unadapted bacteria	2.2	22
Cu-tolerant bacteria	12.0	24
Ni-tolerant bacteria	2.9	88

In general it appears that oxoanions are more toxic than cationic species, probably reflecting the presence of suitable uptake pathways. However, the simple cations, Ag^+, Tl^+, and Hg^{2+}, are especially toxic. It is noteworthy that these most toxic metal cations are all classified as soft acids in chemical character (see section 2.4.2.2). Such cations have a greater affinity for the soft base sites that sulfur centres of proteins provide. The tolerance to iron(III) is very high, as might be expected for a hard ion. High toxicities are associated with species which are neutral molecules at low values of pH; arsenic(III) and (V), molybdenum(VI). It is necessary to obtain consistent results for the toxicities of metal ions towards the range of organisms used in biohydrometallurgical processes before quantitative conclusions may be drawn.

3.6 MIXED CULTURES

Leaching of metals under industrial conditions can occur with the assistance of microorganisms which are naturally present in ores or concentrates. It is not always necessary to inoculate an industrial system with a bacterial culture. The contributions made to the overall process of bacterial oxidation by such naturally present microorganisms, including some of a heterotrophic nature, are usually ignored. This lack of study needs to be rectified since it is possible for the contributions to be beneficial.

3.6.1 The range of microorganisms in ores

The range of microorganisms in ores, and in waters in contact with ores, depends upon their type and the environmental conditions. At low values of pH, bacteria including *T. ferrooxidans*, *T. thiooxidans*, *L. ferrooxidans* and *T. organoparus* (sometimes called *T. acidophilus*), are predominant. Sulfobacilli, other moderate thermophiles and thermoacidophilic archaebacteria predominate in environments which are of a higher temperature. In general these biohydrometallurgically important bacterial species are accompanied by contaminants which include heterotrophic microorganisms, fungi, yeasts and algae. The contaminant species tend to be present as relatively low populations which is indicative that their environments (low pH, low concentrations of organic substances) are far from being optimum for their growth. Serious attention has not been given to the possible roles of these contaminants, but it is quite possible that their presence can be beneficial to the bacterial oxidation process. In general no bacterial oxidations are carried out with absolutely pure strains of chemolithoautotrophic bacteria, but some are carried out beneficially with mixed cultures.

3.6.2 Mechanism of interaction between microorganisms

Currently there are two types of bacterial interaction which may be identified. These are interactions between a chemolithotroph and either (i) another chemolithotroph or (ii) a heterotroph. An example of the first type of interaction is provided by the use of a mixture of *L. ferrooxidans* (an obligate iron oxidizer) and *T. organoparus* (a sulfur oxidizer) in the bacterial oxidation of chalcopyrite (Markosan, 1976). Some results of this work are given in Table 3.6.

Table 3.6 Catalysis of the oxidation of chalcopyrite by *L. ferrooxidans* and *T. organoparus* singly and in association - concentrations of iron(III), copper(II) and sulfur(VI) produced after seven months treatment

organism	concentration/mM		
	Fe(III)	Cu(II)	S(VI)
none	0	1.3	12.4
L. ferrooxidans	0	1.3	12.8
T. organoparus	0	1.7	27.0
both	22.1	12.0	42.8

Used separately the two species were not appreciably effective in solubilizing the chalcopyrite. Their use together was essential for bacterial oxidation to occur. The effectiveness of *T. ferrooxidans*, which is an oxidizer of both iron(II) and sulfur, is not affected by the presence of the sulfur-oxidizer *T. thiooxidans*. If a particular chemolithoautotrophic organism lacks the capacity to oxidize both components (e.g. iron(II) and sulfide ions) of a mineral substrate its performance will be restricted. The admixture of a second organism which ensures the oxidation of both components of the substrate will facilitate the process.

Chemolithoautotrophs produce an effluent of organic substances (exometabolites, section 3.4.1) which could be used as energy and carbon sources for heterotrophic contaminant species. The build-up of some exometabolites may pose a toxicity threat to the bacteria which produced them. A beneficial effect of exometabolites is the complexation of toxic metal ions which reduces their bioavailability. The overall significance of the presence of heterotrophic species in metal leaching systems is not clear mainly because of the lack of suitable data. Some results (Karavaiko 1980) from a study of the interaction of the thermoacidophilic bacterium *S. thermosulfidooxidans* with *L. thermoferrooxidans* and a *Thiobacillus* species are given in Table 3.7.

Table 3.7 The solubilization of iron and arsenic from a mineral substrate by *S. thermosulfidooxidans* in the absence and presence of *L. thermoferrooxidans* and a *Thiobacillus* species

organism(s)	yeast extract	Fe/mM	As/mM
none	-	1.6	1.4
S. thermosulfidooxidans	+	68	9
S. thermosulfidooxidans, Thiobacillus sp. and	+	88	14
L. thermoferrooxidans	-	89	12

The microorganisms were used to solubilize an arsenopyrite concentrate at 45°C. *S. thermosulfidooxidans* is a facultative autotroph which can grow under heterotrophic conditions. Normally these are produced by adding yeast extract in concentrations of around 0.02% w/v of yeast extract added to the medium. The leaching performance of the facultative autotroph is enhanced by the presence of *L. thermoferrooxidans* and the *Thiobacillus* species. The performance of the mixed culture was not diminished by the absence of the yeast extract which might indicate that the facultative autotroph was making use of the exometabolites from the added species. It can be presumed that sulfobacilli obtain organic substances necessary for their growth from the other autotrophic bacterial species *L. thermoferrooxidans*.

Not all heterotrophic species would be expected to enhance the growth and leaching activity of chemolithoautotrophs. Many of them can inhibit the growth of thiobacilli. Only specific heterotrophic bacteria which are capable of oxidizing simple organic substances to water and carbon dioxide are of great interest. These species are called oligotrophic bacteria and help maintain the homeostasis of the chemolithoautotroph.

3.7 ISOLATION OF BACTERIAL CULTURES

The isolation of bacterial cultures which are important to the subject of biohydrometallurgy is carried out by inoculating appropriate nutrient media with samples of ores or acidic solutions (including acid mine drainage). If the conditions are correct, and appropriate microorganisms are present, this leads to the growth of the culture. The choice of medium is important and could determine whether a single or mixed culture is isolated. Such procedures may produce cultures enriched with concomitant organisms. This indicates that the production of pure cultures often presents considerable difficulty. It is usual for natural mixed cultures to be used in bacterial oxidation processes. The separation of components of a mixed culture is carried out for purposes of identification and characterization.

3.7.1 Isolation procedure

Possible sources of bacterial cultures are (i) warm acidic waters from mines or mine dumps (acid mine drainage), (ii) ore minerals and (iii) waters and slurries from hot volcanic springs. The particular cultures of interest are those which survive at low values of pH and which have optimum temperatures for growth in the regions of 30°C (mesophiles), 45°C (moderate thermophiles) or 70-80°C (extreme thermophiles). Samples from the three sources may be treated by a general enrichment procedure in order to encourage the growth of suitable species which may be present.

The general procedure consists of the use of about 10 mL of the water sample, or about one gram of an ore mineral, as an inoculum for 50 mL of one of the recommended medium solutions in a 250 mL Erlenmeyer flask. The flasks and their contents are then incubated at 30°C, 45°C or 80°C until growth is observed. The incubation may be carried out statically in ovens of the appropriate temperature or in thermostated orbital shakers. The latter method allows better oxygenation of the solutions. In some cases the gases supplied to the flasks (normally air) are enriched with carbon dioxide. Bacterial growth may be assessed visually by the appearance of the brown colour typical of iron(III) in the medium in the case of iron oxidizing species. The growth of sulfur oxidizers is monitored by the fall in the pH value of the solution. In all cases the growth should be monitored quantitatively and daily by the instrumental analysis (see Chapter 9 for details) of samples from the flasks. Although there are specific medium solutions which are designed to optimize the growth of particular strains of bacteria it is recommended that, in the cases where no strain has yet been identified, the solution to be used should be a version of the Silverman and Lundgren (1959) 9K medium. The 9K medium consists of two solutions. Solution A is made up by dissolving a mixture consisting of ammonium sulfate (3 g), potassium chloride (0.1 g), dipotassium hydrogen phosphate (0.5 g), magnesium sulfate heptahydrate (0.5 g) and calcium nitrate (0.01 g) in 700 mL of distilled water. Solution B is made up by dissolving iron(II) sulfate heptahydrate (44.2 g) and 1 mL of 5 M sulfuric acid in 300 mL of distilled water. The solutions are sterilized separately, solution A at atmospheric pressure, solution B at 0.5 atm. The solutions are then mixed before use. Sterilization, although a proper microbiological procedure, is not strictly necessary for materials used for the growth of acidophilic chemolithotrophs since species not of such a class will not survive in the medium. For the promotion of the growth of iron oxidizing species it is normal to omit the iron(II) sulfate from the medium and to supply any suitable bacteria with finely ground pyrite (1% w/v) as a source of iron. The form of the 9K medium in such a case would be solution A to which the sulfuric acid and an extra 300 mL of distilled water have been added. To encourage the growth of sulfur oxidizing species elemental sulfur is added to the extent of 10 g per litre of medium solution. Some moderate and all extreme thermophiles are facultative autotrophs with growth rates which are enhanced in the presence of either yeast extract (0.02% w/v) or a sugar such as glucose.

The initial procedure with any particular sample suspected of containing useful bacteria is dependent upon the source of the material (solid or solution). Mesophiles and moderate thermophiles possess temperature profiles for their growth which overlap considerably. They occur together in some sources. There is practically no overlap between the temperature profiles of the mesophiles and moderate thermophiles with the extreme thermophiles so different procedures are carried out with the two groups of organisms.

3.7.1.1 Mesophiles and moderate thermophiles

If the temperature of the water source is around 30°C it is most unlikely that extreme thermophiles are present in the sample. Mesophiles tend to be obligate chemolithoautotrophs while moderate thermophiles are usually facultative chemolithoautotrophs. Any contaminants which may be present could be heterotrophic. To encourage the growth of required species it is necessary to arrange suitable conditions. Duplicate sets of two medium solutions should be used, one solution consisting of the modified (iron(II)-free) 9K medium plus 1% w/v pyrite, the other being made up from the modified 9K medium containing pyrite and 0.02% w/v of yeast extract. One set should be incubated at 30°C to encourage the growth of any mesophiles, the other being incubated at 45°C to facilitate the growth of any moderate thermophiles.

3.7.1.2 Extreme thermophiles

Potential sources of extreme thermophilic cultures originate in and around hot volcanic springs where water temperatures are between 70-80°C. Since extreme thermophiles are either sulfur oxidizers (e.g. *Sulfolobus acidocaldarius*) or sulfur and iron/sulfide mineral oxidizers (e.g. *Acidanus brierleyi*), and are facultative chemolithoautotrophs, two solutions should be incubated at 80°C. These should be made up from the 9K medium (iron(II)-free) with either sulfur or pyrite present.

3.7.2 Subculturing procedure

During the isolation procedures described above the substrates become depleted as growth occurs. This depletion is monitored and when it has reached a value in excess of 50% a subculturing procedure is normally carried out. Half of the contents of a flask are used as the inoculum for a fresh medium with appropriate additions of pyrite, sulfur and yeast extract. The remaining half is usually disposed of or can be used to double up on the procedure. The subculturing is continued until the cell count approaches 10^{8-9} cells mL^{-1}.

3.7.3 Isolation of pure cultures

If cell growth has taken place, as indicated by the usage of the supplied substrate(s), it is usually possible to make an assessment of the purity of the culture by microscopy. Experienced observers are able to make a reasonably accurate assessment by the direct microscopy of a drop of the solution on a microscope slide, using magnifications of around x1000. If the culture does contain more than one species it is possible, in principle, to separate them by a plating method. After incubation of the inoculated plates colonies of iron oxidizing bacteria appear as brown spots. However, the application of these techniques to autotrophic bacteria is not straightforward. There are difficulties associated with the production of solid media which have high enough acidity to promote the growth of acidophilic bacteria. Furthermore, the high acidity may cause the hydrolysis of some commercial gelling agents and the liberation of inhibitory compounds (principally carbohydrates). Growth of bacteria on such media at higher pH values tends to produce yellow/brown iron(III) precipitates that mask the presence of bacterial colonies. Several solutions to these problems have been proposed, including the use of silica gel plates or those consisting of polyacrylamide gels. The more successful approaches appear to be those that involve two gel layers (Harrison, 1984). Johnson and McGinness (1991) have developed a two layer gel in which the lower layer contained an heterotrophic acidophilic organism which utilized any carbohydrates released in the upper layer. This prevents the inhibition of the growth of the autotroph in the upper layer. When colonies of autotrophic bacteria are established individual pure colonies may then be used to inoculate small volumes (5 mL) of liquid media containing either iron(II) or pyrite and are allowed to grow. The properties of the component species of the mixture can then be studied separately and grown in large quantities if necessary.

3.7.4 Maintenance and storage of cultures

It is advisable to maintain a store of any useful cultures which have been isolated. This is easily done by the subculturing methods outlined above. It is possible to maintain a static culture at lower than optimum temperatures for considerable periods of time before the nutrients and substrates require replenishment. It is important to preserve the unadapted strain of a culture so that it may be adapted to new conditions rather than to adapt a culture to entirely new conditions. It is also important to have a back-up culture of any adapted strains for use in replacements should the one being used is accidentally lost.

3.8 REFERENCES

Bagdigian, R.M. & Myerson, A.S. (1986) *Biotechnol. Bioeng.*, **28**, 467.
Barrett, J. (1992) *unpublished work*.
Beijerinck, M.W. (1904) *Zentralbl. Bakteriol. Parasitenkd. Infektionskr. Hyg. Abt.*, **11**, 593.
Berry, V.K. & Murr, L.E. (1976) *34th Ann. Proc. Electron Microscopy Miami Beach, Florida, 1976*, Bailey, G.W. (ed.) p.132.
Brierley, C.L. & Brierley, J.A. (1982) *Zentralbl. Bakteriol. Parasitenkd. Infektionskr. Hyg. Abt.*, **10**, 289.
Brierley, J.A. (1978) *Appl. Environ. Microbiol.*, **36**, 523.
Brierley, J.A. & Le Roux, N.W. (1977) *Conference Bacterial Leaching*, Schwartz, W., (ed.) GBF Monograph No.4 Weinheim, New York, Verlag Chemie, p.55.
Brierley, J.A. & Lockwood, S.T. (1977) *FEMS Microbiol. Lett.*, **2**, 163.
Brock, T.D. & Gustafson, J. (1976) *Appl. Environ. Microbiol.*, **32**, 567.
Cook, T.M. (1964) *J. Bacteriol.*, **88**, 620.
Dispirito, A.A., Dugan, P.R. & Tuovinen, O.H. (1981) *Biotechnol. Bioeng.*, **23**, 2761.
Dispirito, A.A., Dugan, P.R. & Tuovinen, O.H. (1983) *ibid.*, **25**, 1163.
Golovacheva, R.S., Golyshina, O.V., Karavaiko, G.I., Dorofeyev, A.G., Pivovarova, T.A. & Chernych, N.A. (1992) *in press*.
Golovacheva, R.S. & Karavaiko, G.I. (1977) *Second International Symposium on Microbial Growth on C_1 Compounds, Pushchino, USSR Academy of Sciences, 1977*, p.108.
Gormely, L.S. & Duncan, D.W. (1974) *Can. J. Microbiol.*, **20**, 1453.
Guay, R. & Silver, M. (1975) *ibid.*, **21**, 281.
Harrison, A.P.Jnr., Jarvis, B.W. & Johnson, J.L. (1980) *J. Bacteriol.*, **143**, 448.
Harrison, A.P. (1984) *Ann. Rev. Microbiol.*, **38**, 265.
Harrison, A.P.Jnr. & Norris, P.R. (1985) *FEMS Microbiol. Lett.*, **30**, 99.
Hendy, N.A. (1987) *J. Industr. Microbiol.*, **1**, 389.
Johnson, D.B. & McGinness, S. (1991) *J. Microbiol. Methods* **13**, 113.

Johnson, D.B., Said, M.F., Ghauri, M.A. & McGinness, S. (1989) *Biohydrometallurgy-89, International Symposium, Jackson Hole, Wyoming, 1989.* Salley, J., McCready, R.G.L. & Wichlacz, P.L. (eds) CANMET SP89-10, p.403.
Karavaiko, G.I. (1980) *unpublished work.*
Karavaiko, G.I., Golovacheva, R.S., Pivovarova, T.A., Tzaplina, I.A. & Vartanjan, N.S. (1988) *Biohydrometallurgy-87, International Symposium, Warwick, U.K. 1987,* Norris, P.R. & Kelly, D.P. (eds) Science and Technology Letters, p.29.
Kovalenko, E.V. & Malakhova, P.T. (1983) *Mikrobiologiya* **52**, 962.
Le Roux, N.W., Wakerley, D.S. & Hunt, S.D. (1977) *J. Gen. Microbiol.*, **100**, 197.
MacIntosh, M.E. (1978) *J. Gen. Microbiol.*, **105**, 215.
Markosjan, G.E. (1972) *Biol. Zh. Armenii* **25**, 26.
Markosjan, G.E. (1976) *Ecology and Geochemical Activity of Microorganisms, Akad.Nauk USSR* Pushchino, p.160.
Marsh, R.M. & Norris, P.R. (1983) *Biotechnol. Lett.*, **5**, 585.
Mergeay, M. (1991) *Tibtech.*, **9**, 17.
Natarajan, K.A. & Iwasaki, I. (1983) *Hydrometallurgy* **10**, 329.
Natarajan, K.A., Iwasaki, I. & Reid, K.J. (1983) *Recent Progress in Biohydrometallurgy*, Rossi, G. & Torma, A.E. (eds) Cagliari, p.169.
Nobar, A.M., Ewart, D.K., Alsaffar, L., Barrett, J., Hughes, M.N. & Poole, R.K. (1988) *Biohydrometallurgy - 87, International Symposium, Warwick, U.K. 1987,* Norris, P.R. & Kelly, D.P. (eds) Science and Technology Letters, p.530.
Norris, P.R., Brierley, J.A. & Kelly, D.P. (1980) *FEMS Microbiol. Lett.*, **7**, 119.
Norris, P.R. & Kelly, D.P. (1978) *Metallurgical Applications of Bacterial Leaching and Related Microbiological Phenomena*, Murr, L.E., Torma, A.E. & Brierley, J.A. (eds) Academic Press, p.443.
Pronk, J.T., Meijeer, W.M., Hazeu, W., van Dijken, J.P., Bos, P. & Kuenen, J.G. (1991) *Appl. Environ. Microbiol.*, **57**, 2057.
Rossi, G. (1990a) *Biohydrometallurgy.* McGraw-Hill, p 200.
Rossi, G. (1990b) *ibid.*, p.270.
Schlegel, H.G. (1986) *General Microbiology*, 6th ed., Cambridge University Press.
Silverman, M.P. & Lundgren, D.G. (1959) *J. Bacteriol.*, **77**, 642.
Stevens, C.J., Dugan, P.R. & Tuovinen, O.H. (1986) *Biotechnol. Appl. Biochem.*, **8**, 351.
Sugio, T., Domatsu, C., Munakata, O., Tano, T & Imai, K. (1985) *Appl. Environ. Microbiol.*, **49**, 1401.
Tributsch, H. & Bennett, J.C. (1981a) *J. Chem. Technol. Biotechnol.*, **31**, 565.
Tributsch, H. & Bennett, J.C. (1981b) *ibid.*, p.627.
Vitaya, V.B. & Toda, K. (1991) *Biotechnol. Progr.*, **7**, 427.
Wakao, N., Mishima, M., Sakurai, Y. & Shiota, H. (1984) *J. Gen. Appl. Microbiol.*, **30**, 63.
Weiss, R.L. (1973) *J. Gen. Microbiol.*, **77**, 501.
Wood, A.P & Kelly, D.P. (1983) *FEMS Microbiol. Lett.*, **20**, 107.
Yakontova, L.K. (1985) *Biogeotechnology of Metals*, Karavaiko, G.I. & Groudev, S.N. (eds) Centre of International Projects, GKNT, Moscow, p.216.
Zavarsin, G.A. (1972) *Mikrobiologiya* **41**, 369.

4

The chemistry of bacterial oxidation reactions

4.1 INTRODUCTION

The three topics which are important in the understanding of the chemistry of bacterial oxidation reactions and the environmentally acceptable disposal of waste solutions and solids are (i) the measurement of pH, (ii) the speciation of the components of the solutions used in bacterial oxidation systems and (iii) electrode potential measurements. The chemistry of waste disposal is dealt with fully in Chapter 7.

The chemistry of bacterial oxidation reactions may be understood by taking into account the speciation of all the components of a system and any interactions between the components. This involves the consideration of known equilibrium constants at appropriate temperatures for acid dissociations and complex formations to determine the different species (ions or neutral molecules) which occur in solution and their concentrations. The results of speciation calculations (or experimental determinations) are usually expressed in terms of alpha (α) values which represent the fractions of the species in which the oxidation state of a particular element exist.

The speciation of any system is normally highly dependent upon the pH value of the solution and it is therefore important to understand the relationship between the theoretical basis of pH and its practical application.

A very convenient method of summarizing the thermodynamic characteristics of a reaction is to use the change in standard Gibbs energy per mole of electrons in the form of the standard electrode potential, $E^{\ominus}$ ($= -\Delta G^{\ominus}/nF$, where n is the number of electrons transferred from the reducing agent to the oxidizing agent and F is the Faraday, 96484.56 Coulomb mol^{-1}, one volt-Coulomb (1 V.C) being equal to one Joule (1 J)). To be thermodynamically feasible a reaction must have a negative $\Delta G^{\ominus}$ value and, in consequence, the corresponding $E^{\ominus}$ value must be positive.

The variation of E with pH for any half-reaction may be represented by the appropriate Nernst equation (Atkins, 1982) and is very important in the recognition of reaction feasibility at pH values other than zero. When values of E or $E^{\ominus}$ (standard reduction potentials of half-reactions on the hydrogen scale) are used it should be understood that they are convenient expressions of the thermodynamic characteristics of a system and do not represent any implication as to the 'galvanic' nature of the reactions involved. Just because an overall reaction may be split into two halves, both of which involve free electrons, it does not mean that they have any relevance to the mechanism of the reaction.

The solubilization of metals, from their ore minerals, or the liberation of encapsulated gold particles from sulfidic or arsenosulfidic minerals, by bacterial oxidation, takes place under aqueous conditions at atmospheric pressure and at temperatures between 30-80°C. The pH value of the aqueous leaching solution is normally in the range 0.5-1.5. It is in this range of pH values that the speciation of solubilized products of the bacterial oxidation reactions are of particular interest. The understanding of the speciation also has important consequences for the treatment of aqueous effluent from the plant (treated in detail in Chapter 7).

4.2 ASPECTS OF pH MEASUREMENT

Section 4.3 is devoted to speciation, which deals with the variation of the fractions of various species with the pH of the solution. It is important to describe the relationship between pH and the hydrogen ion concentration, [H$^+$], the latter quantity being that used in the calculations.

The theoretical definition of pH is given by the equation:

$$\text{pH} = -\log_{10} a_{(\text{H}^+)} \tag{4.1}$$

where $a_{(\text{H}^+)}$ represents the *activity* of the hydrogen ion. The activity, a, of any component of a solution is related to its concentration by the equation:

$$a = (m/m^\ominus)\gamma \tag{4.2}$$

where m is the molality (the number of moles per kilogram of solvent) of the component, $m^\ominus$ its standard molality (1 mol kg^{-1}) and γ is its *activity coefficient*. The activity coefficient is a factor which takes into account the deviation of the component from ideal behaviour because of the interfering presence of a non-zero concentration of identical and other ions. As the molality approaches zero (conditions where ideal behaviour is to be expected because the ions are sufficiently far apart that they do not interact with each other) the value of the activity coefficient approaches unity. Although the strict definition of pH involves the molality of the hydrogen ion, in practice it is normal to use its molarity (the number of moles per litre of solution) to express concentration. There is only a significant difference between molality and molarity at exceptionally high concentrations which do not apply to any of the solutions used in bacterial oxidation reactions. The molarity of a solution is slightly dependent upon its temperature but the molality is not. The major difficulty with the theoretical definition of pH is that, although the molarity of the hydrated proton may be obtained by titration with a suitable base, its activity coefficient is impossible to determine experimentally. This statement applies to any *ionic* species because it is impossible for a solution of any ion not to contain a counter ion of the opposite charge. In a solution consisting of two oppositely charged ions the individual ionic activity coefficients, γ_+ and γ_-, cannot be determined. The parameter which can be determined experimentally is the *mean ionic activity coefficient* which is defined (for a 1:1 electrolyte) as:

$$\gamma_\pm = (\gamma_+ \cdot \gamma_-)^{1/2} \qquad (4.3)$$

The value of the mean ionic activity coefficient approaches unity as the molality (or the molarity) of the electrolyte approaches zero.

Activity coefficients of individual ions may be estimated theoretically from Debye-Hückel theory and its extensions. Their values are dependent upon the *ionic strength* of the solution. The ionic strength of a solution is defined by the equation:

$$I = \tfrac{1}{2} \sum_i m_i z_i^2 \qquad (4.4)$$

for all the component ions, i, of the solution, m_i representing the molality and z_i the charge of the ion, i. The value of the activity coefficient of an ion is given by an extension of the theoretical Debye-Hückel equation:

$$-\log_{10}\gamma_i = A z_i^2 I^{1/2}/(1 + BI^{1/2}) \qquad (4.5)$$

where A is a constant with the value 0.5091 at 25°C, z_i is the ionic charge and B is a parameter having a value usually taken to be 1 (L mol^{-1})$^{1/2}$. The equation is reasonably accurate at values of ionic strength up to about 0.1 M but is not reliable at higher values. With careful measurements of appropriate cell potentials it is possible to determine the mean activity coefficient for aqueous hydrochloric and sulfuric acids at various ionic strengths. For instance, the values of $\gamma_\pm$ for the two acids at a total ionic strength of one molal are 0.809 and 0.131 respectively at 25°C and 0.796 and 0.266 at 0.1 molal (Glasstone, 1947). It is obvious from the large differences between the two values that the counter ions to the proton, Cl$^-$ and HSO$_4^-$ respectively in the two cases, have a large influence upon the value of $\gamma_\pm$. At 0.001 molal the $\gamma_\pm$ value for HCl is 0.996 implying that the solution is sufficiently dilute for its behaviour to be regarded as practically ideal. The equivalent value for sulfuric acid is 0.83 so the solution is still far from behaving in an ideal manner. The relatively complex nature of the solutions used in bacterial oxidation processes ensures that the measurement of pH must have a firm experimental basis which is not dependent upon immeasurable quantities.

From the above discussion of pH measurement and the nature of the relationship between pH and the hydrogen ion concentration it is clear that there must be some doubt about the conversion of pH values to hydrogen ion concentrations which would give an accurate indication of the speciation fractions (the 'alpha' values of section 4.3) for a bacterial oxidation system. The main problem is to obtain an accurate value for the hydrogen ion concentration for a given pH. The calculations of the alpha values in the next section are based upon the theoretical relationship between pH and [H$^+$] (equation (4.1)) with some estimated values of γ. The exclusion (or inclusion!) of such values can lead to an error range of about ±0.4 on the pH scale. In laboratory and plant practice it is important to use generally accepted standard solutions to calibrate the pH meters used. It is by doing this that consistency meeting with general acceptability can be achieved. This is important to allow different systems to be compared in a reasonably scientific manner. The practical aspects of pH measurement are dealt with in Chapter 9.

4.3 SPECIATION

The aqueous oxidation states of elements whose speciation is important in the understanding of the chemistry of bacterial oxidation processes are (i) sulfur(VI), (ii) arsenic(III), (iii) arsenic(V), (iv) iron(II), (v) iron(III) and (vi) ions of any other metals solubilized. In all cases it is essential to take into account any ions produced by hydrolysis or by complex formation. In the following discussion the concentrations of the components of a system are expressed by the conventional method of enclosing the formula in square brackets. The strict application of thermodynamics would involve the use of *activities* instead of concentrations but this depends upon the knowledge of *activity coefficients* for every component of a system. These data are not generally available for the systems used in bacterial oxidation processes so the following discussions must be regarded as approximations to the real state of affairs. There are methods of estimating ionic activity coefficients even at the high ionic strengths used in bacterial oxidation solutions. They are subject to considerable uncertainties and it is considered that their inclusion would possibly introduce more errors than they would potentially correct. The calculations make use of concentrations (i.e. equilibrium constants are used as conditional values) throughout and assume that activity coefficients have the value of unity except for cases where reasonable estimates are available.

4.3.1 Speciation of sulfur(VI) in sulfuric acid solutions

The speciation of sulfur(VI) in sulfuric acid solutions involves only the relative amounts of the hydrogen sulfate, HSO_4^-, and sulfate, SO_4^{2-}, ions since the sulfuric acid molecule, H_2SO_4, is totally ionized in any aqueous solution. If the total analytical concentration of sulfur(VI) in a system is denoted by T(S), then an equation representing the speciation of sulfur(VI) ions may be written as:

$$T(S) = [HSO_4^-] + [SO_4^{2-}] \tag{4.6}$$

The equilibrium between the two sulfur(VI) ions and aquated protons, $H_3O^+(aq)$, (represented by H$^+$ in the remainder of this book) is represented by the equation:

$$HSO_4^- \rightleftharpoons H^+ + SO_4^{2-} \qquad (4.7)$$

with the equilibrium constant being given by:

$$K_{S2} = [H^+][SO_4^{2-}]/[HSO_4^-] \qquad (4.8)$$

where K_{S2} is the second dissociation constant of sulfuric acid having a value of 0.012 (pK_2 = 1.92) at 25°C (Perrin, 1969a). The ratio of the concentrations of the two ions, $[HSO_4^-]/[SO_4^{2-}]$, is given by $[H^+]/K_{S2}$. The fraction of sulfur(VI) which is in the form of the hydrogen sulfate ion, α_{HS}, is given by:

$$\alpha_{HS} = [HSO_4^-]/T(S) \qquad (4.9)$$

and its reciprocal by:

$$1/\alpha_{HS} = T(S)/[HSO_4^-] = 1 + [SO_4^{2-}]/[HSO_4^-] = 1 + K_{S2}/[H^+] \qquad (4.10)$$

This allows α_{HS} to be written as:

$$\alpha_{HS} = (1 + K_{S2}/[H^+])^{-1} = [H^+]/([H^+] + K_{S2}) \qquad (4.11)$$

The fraction of the sulfur(VI) present as the sulfate ion, α_S, is thus given by:

$$\alpha_S = 1 - \alpha_{HS} = K_{S2}/([H^+] + K_{S2}) \qquad (4.12)$$

The variation of α_{HS} and α_S as a function of pH is shown in Fig. 4.1 for the range of pH values zero to four. In the pH range 0-2, which is appropriate for bacterial oxidation systems, the hydrogen sulfate ion is the predominant sulfur(VI) species. In a system where the sulfur(VI) species are involved in the formation of complexes the values of α_S and α_{HS} are smaller but their ratio would be unaltered at any particular pH value.

4.3.2 Arsenic(III)

The molecular formula of arsenic(III) acid may be represented by the formula H_3AsO_3 (or as $As(OH)_3$). The first (and only important) dissociation constant is 6.607 x 10^{-10} at 25°C (the corresponding pK value is 9.18, Perrin, 1969b). This means that, at a pH value of 9.18, the concentrations of the neutral molecule and the arsenate(III) ion, $H_2AsO_3^-$, are equal. At pH values below 9.18 the neutral molecule predominates and in the pH range 0.5-1.5, which is appropriate for bacterial oxidation reactions, the arsenate(III) ion plays virtually no part. At pH values below 0.3 the protonated form of the acid, $H_4AsO_3^+$, becomes predominant (Crecelius et al. 1986).

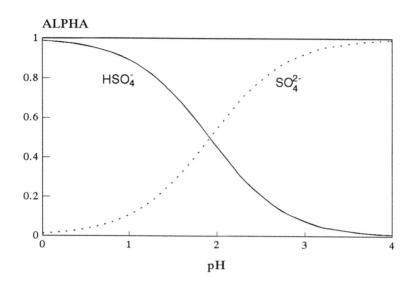

Fig. 4.1 Speciation of sulfur(VI) in aqueous solution

4.3.3 Arsenic(V)

Arsenic(V) acid, H_3AsO_4, possesses three dissociable protons. The three dissociation equilibria may be written as:

$$H_3AsO_4 \leftrightarrows H^+ + H_2AsO_4^- \qquad (4.13)$$

$$H_2AsO_4^- \leftrightarrows H^+ + HAsO_4^{2-} \qquad (4.14)$$

$$HAsO_4^{2-} \leftrightarrows H^+ + AsO_4^{3-} \qquad (4.15)$$

The equations for the equilibrium (acid dissociation) constants are given below, the values of K and pK being those at 25°C (Perrin, 1969c):

$$K_1 = [H^+][H_2AsO_4^-]/[H_3AsO_4] = 6.457 \times 10^{-3} \ (pK_1 = 2.19) \qquad (4.16)$$

$$K_2 = [H^+][HAsO_4^{2-}]/[H_2AsO_4^-] = 1.148 \times 10^{-7} \ (pK_2 = 6.94) \qquad (4.17)$$

$$K_3 = [H^+][AsO_4^{3-}]/[HAsO_4^{2-}] = 3.162 \times 10^{-12} \ (pK_3 = 11.5) \qquad (4.18)$$

The total arsenic(V) concentration, T(As), may be written as:

$$T(As) = [H_3AsO_4] + [H_2AsO_4^-] + [HAsO_4^{2-}] + [AsO_4^{3-}] \qquad (4.19)$$

Speciation

The reciprocal of the fraction existing as the neutral molecule, α_3 (the subscript indicating the number of undissociated protons), is obtained by dividing equation (4.19) by the concentration of neutral arsenic(V) acid, $[H_3AsO_4]$, which gives the equation:

$$1/\alpha_3 = 1 + \frac{[H_2AsO_4^-]}{[H_3AsO_4]} + \frac{[HAsO_4^{2-}]}{[H_3AsO_4]} + \frac{[AsO_4^{3-}]}{[H_3AsO_4]} \tag{4.20}$$

Equation (4.16) for the first dissociation allows the second term on the right-hand-side to be replaced by $K_1/[H^+]$. A combination of equations (4.16) and (4.17) gives $K_1K_2/[H^+]^2$ for the third term and a combination of all three dissociation equations gives $K_1K_2K_3/[H^+]^3$ for the fourth term. After substituting these expressions, equation (4.20), in a reciprocal form, gives the value of α_3 as:

$$\alpha_3 = \left[1 + \frac{K_1}{[H^+]} + \frac{K_1K_2}{[H^+]^2} + \frac{K_1K_2K_3}{[H^+]^3}\right]^{-1} \tag{4.21}$$

The equations which give α_2, α_1 and α_0 (the fractions of the ions, $H_2AsO_4^-$, $HAsO_4^{2-}$ and AsO_4^{3-}, respectively) are dependent upon the value of α_3, which may be written in the form:

$$\alpha_3 = [H_3AsO_4]/T(As) \tag{4.22}$$

The value of α_2 is given by:

$$\alpha_2 = [H_2AsO_4^-]/T(As) \tag{4.23}$$

which combined with equation (4.16) gives:

$$\alpha_2 = \frac{[H_3AsO_4].K_1}{T(As)[H^+]} = \alpha_3 \cdot \frac{K_1}{[H^+]} \tag{4.24}$$

The use of equations (4.17) and (4.18) allows similar calculations to give the values of α_1 and α_0 as:

$$\alpha_1 = \alpha_3 \cdot \frac{K_1 K_2}{[H^+]^2} \qquad (4.25)$$

$$\alpha_0 = \alpha_3 \cdot \frac{K_1 K_2 K_3}{[H^+]^3} \qquad (4.26)$$

The values of the four α's are plotted against pH in Fig. 4.2 over the range of pH values from zero to fourteen.

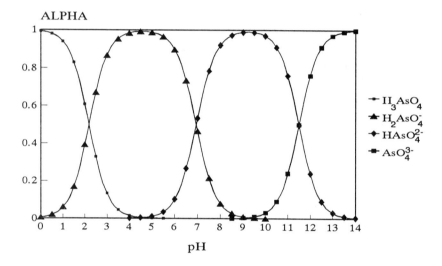

Fig. 4.2 Speciation of arsenic(V) in aqueous solution

The results indicate that the neutral arsenic(V) acid molecule, H_3AsO_4, is the predominant species in the pH range 0-2. The ionic As(V) species are only important at higher pH values when arsenic is precipitated from the liquid effluent (Chapter 7).

4.3.4 Iron(II)

Iron(II) in aqueous solution may exist as the hexaaquairon(II) ion, $[Fe(H_2O)_6]^{2+}$, but at high pH values this ion undergoes hydrolysis to the hydroxopentaaquairon(II) ion, $[Fe(H_2O)_5OH]^+$. The equilibrium constant for the hydrolysis is 1.81×10^{-7} (pK = 6.74) at 25°C (Perrin, 1969d):

$$[Fe(H_2O)_6]^{2+} \rightleftarrows [Fe(H_2O)_5OH]^+ + H^+ \qquad (4.27)$$

This hydrolysis is of no importance to bacterial oxidation reactions taking place within their normal pH range. Any iron(II) in a bacterial oxidation system is present entirely as the hexa-hydrated doubly charged cation providing that no complexing ligands are present. There are no reported sulfur(VI) or arsenic(V) complexes of iron(II).

4.3.5 Iron(III)

Iron(III) disposal is a feature of many industrial processes and as a consequence its hydrolysis has been the subject of a large number of investigations. Flynn (1984) has published a valuable and comprehensive review of the literature up to that date. The proceedings of a symposium (Dutrizac and Monhemius, 1986) contain forty one papers which include discussions of the problems associated with iron control in hydrometallurgy.

The unhydrolyzed iron(III) ion exists in aqueous solution and the crystalline state as the hexaaquairon(III) ion, $[Fe(H_2O)_6]^{3+}$. In the crystalline state, with either nitrate ion or perchlorate ion as the counter anion, the hexaaquairon(III) ion is responsible for the observed pale violet colour of the solids. In aqueous solution the hydrated ion undergoes hydrolysis, even at pH values as low as two, to give red/brown coloured species. Major hydrolysis products which have been definitely identified are the ions $[Fe(H_2O)_5OH]^{2+}$, $[(H_2O)_5FeOFe(H_2O)_5]^{4+}$ and $[Fe(H_2O)_4(OH)_2]^+$. At higher pH values the system becomes more complex with the production of a variety of polymeric species and colloidal particles leading, as the pH increases, to the precipitation of solid materials of variable composition. At pH values greater than ten the complex $[Fe(OH)_4]^-$ contributes to an increased solubility of iron(III). In addition to hydrolysis complex ion formation occurs in the presence of ligands such as sulfur(VI) or arsenic(V) species.

As iron(III) is a major participant in bacterial oxidation systems it is important to treat its speciation fully. This may be done in stages by considering three systems of increasing complexity:

(i) the iron(III)-perchlorate system in which the iron(III) undergoes hydrolysis but no other complexation occurs,

(ii) the iron(III)-sulfur(VI) system in which iron(III)-sulfate and iron(III)-hydrogen sulfate complexes occur in addition to hydrolysis, and

(iii) the iron(III)-sulfur(VI)-arsenic(V) system in which iron(III)-sulfate, iron(III)-hydrogen sulfate and iron(III)-arsenic(V) complexation occurs, in addition to hydrolysis.

4.3.5.1 Speciation of iron(III) in the presence of perchlorate ion.

The perchlorate ion is chosen as the counter ion to the iron(III) because it is considered to be non-complexing. The well characterized equilibria which represent the main hydrolysis products of iron(III) in such a system are:

$$[Fe(H_2O)_6]^{3+} \rightleftarrows H^+ + [Fe(H_2O)_5OH]^{2+} \qquad (4.28)$$

$$[Fe(H_2O)_5OH]^{2+} \rightleftarrows H^+ + [Fe(H_2O)_4(OH)_2]^+ \qquad (4.29)$$

$$2[Fe(H_2O)_5OH]^{2+} \rightleftarrows H_2O + [(H_2O)_5FeOFe(H_2O)_5]^{4+} \qquad (4.30)$$

$$[Fe(H_2O)_4(OH)_2]^+ \rightleftarrows H^+ + [Fe(H_2O)_3(OH)_3] \qquad (4.31)$$

$$[Fe(H_2O)_3(OH)_3] \rightleftarrows H^+ + [Fe(H_2O)_2(OH)_4]^- \qquad (4.32)$$

The ions $[Fe_3(OH)_4]^{5+}$ and $[Fe_{12}(OH)_{34}]^{2+}$, which have been reported but whose characterizations are not very certain, are ignored for the purposes of the subsequent calculation which is given as an example. There has been some doubt about the nature of the bridging in the diamagnetic dimeric ion which is now regarded (Cotton and Wilkinson, 1988) as having a linear Fe-O-Fe bridge so allowing it to be formulated as $[(H_2O)_5FeOFe(H_2O)_5]^{4+}$. Previous representations have formulated it as having a double hydroxo-bridge, $Fe(OH)_2Fe$. The neutral hydrolysis product, $[Fe(OH)_3)]$, (omitting the three water molecules) represents the precursor of various colloidal particles and the solid 'iron(III) hydroxide' which is produced at low acid concentrations. The nature of this 'hydroxide' is discussed further in Chapter 7 and is best represented either as FeO(OH) (iron(III) oxohydroxide, the chemical name for the mineral goethite which has that composition) or as a hydrated iron(III) oxide, $Fe_2O_3.xH_2O$.

In the ensuing discussions of speciation the conventional representations of the above complex ions are dispensed with. To simplify the algebra which follows the relevant species are referred to by using the symbolism, Fe(p,q), where p represents the number of iron(III) centres and q represents the number of water molecules which have theoretically undergone hydrolysis (usually q is the number of OH groups).

The equilibrium constants which are used in the calculation are defined (in terms of concentrations) as follows:

$$K_1 = [Fe(1,1)][H^+]/[Fe(1,0)] \qquad (4.33)$$

$$K_2 = [Fe(1,2)][H^+]/[Fe(1,1)] \qquad (4.34)$$

$$K_3 = [Fe(1,3)][H^+]/[Fe(1,2)] \qquad (4.35)$$

$$K_{22} = [Fe(2,2)]/[Fe(1,1)]^2 \qquad (4.36)$$

$$K_4 = [Fe(1,4)][H^+]/[Fe(1,3)] \quad (4.37)$$

Although equation (4.36) does not contain an aquated proton term the formation of the Fe(2,2) dimer is dependent upon pH because of the presence of the $[Fe(1,1)]^2$ term. The five equations (4.33-4.37) contain seven 'unknowns'. It is not necessary to calculate the value of the hydrogen ion concentration so that only one more equation is needed to effect a solution of the problem which is to calculate the fractions of the various species present at a given value of pH. The required equation is that which represents the total iron(III) concentration, T(Fe), which may be written as:

$$T(Fe) = [Fe(1,0)] + [Fe(1,1)] + [Fe(1,2)] + [Fe(1,3)]$$
$$+ 2[Fe(2,2)] + [Fe(1,4)] \quad (4.38)$$

Equation (4.38) can be reduced to one containing the two unknowns, $[H^+]$ and $[Fe(1,0)]$, by substitution of the other unknowns by their values given by equations (4.33-4.37). The equation then becomes:

$$T(Fe) = [Fe(1,0)] + K_1[Fe(1,0)]/[H^+] + K_1K_2[Fe(1,0)]/[H^+]^2 + K_1K_2K_3[Fe(1,0)]/[H^+]^3$$
$$+ 2K_{22}K_1^2[Fe(1,0)]^2/[H^+]^2 + K_1K_2K_3K_4[Fe(1,0)]/[H^+]^4 \quad (4.39)$$

Equation (4.39) is a quadratic in $[Fe(1,0)]$ which may be represented by the further abbreviation, F, to give the simplified and rearranged form:

$$2K_{22}K_1^2F^2/[H^+]^2 + F\cdot\{1 + K_1/[H^+] + \beta_2/[H^+]^2$$
$$+ \beta_3/[H^+]^3 + \beta_4/[H^+]^4\} - T(Fe) = 0 \quad (4.40)$$

where the β's, following normal convention (β represents an overall formation constant, the K values being step-wise formation constants), are given by $\beta_2 = K_1K_2$, $\beta_3 = K_1K_2K_3$ and $\beta_4 = K_1K_2K_3K_4$. Equation (4.40) may be solved for F, by using the formula for quadratics (Hall and Knight, 1948), for any values of $[H^+]$ and T(Fe):

$$F = (-b \pm (b^2-4ac)^{1/2})/2a \quad (4.41)$$

where a is the coefficient of F^2 in the first term, b is the coefficient of F in the second term, and c is equal to -T(Fe). Each term in equation (4.40) represents the concentration of a particular species and when divided by T(Fe) gives the appropriate alpha value. The fraction, denoted by α_{10}, of iron(III) present as the free ion, Fe(1,0), is given by:

$$\alpha_{10} = F/T(Fe) \tag{4.42}$$

The fractions present as Fe(1,1) (α_{11}), Fe(1,2) (α_{12}), Fe(1,3) (α_{13}), Fe(2,2) (α_{22}) and Fe(1,4) (α_{14}) are given by the following equations:

$$\alpha_{11} = \alpha_{10}K_1/[H^+] \tag{4.43}$$

$$\alpha_{12} = \alpha_{10}\beta_2/[H^+]^2 \tag{4.44}$$

$$\alpha_{13} = \alpha_{10}\beta_3/[H^+]^3 \tag{4.45}$$

$$\alpha_{22} = 2\alpha_{10}^2 K_1^2 K_{22}.T(Fe)/[H^+]^2 \tag{4.46}$$

$$\alpha_{14} = \alpha_{10}\beta_4/[H^+]^4 \tag{4.47}$$

Plots of these fractions as a function of pH are shown in Fig. 4.3 for a total iron(III) concentration of 0.3 M over the pH range from zero to six. The speciation of iron(III) is dependent upon its total concentration because of the inclusion of the Fe(2,2) complex which introduces a quadratic term into the algebra.

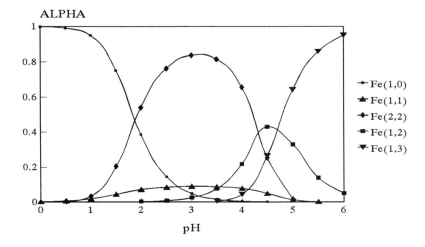

Fig. 4.3 Speciation of 0.3 M iron(III) with perchlorate ion as the counter ion

The values of the hydrolysis constants used in the calculations are those recommended by Baes and Mesmer (1976) and which include estimates of the mean ionic activity coefficients. In the pH range 0-2 the unhydrolyzed Fe(1,0) ion predominates with a minor contributions from the hydrolysis products, Fe(2,2) and Fe(1,1). The other hydrolysis products occur to some extent in the pH range two to six but the main feature of the higher pH range is the eventual predominance of the Fe(1,3) neutral species which is formed as the precursor of the colloidal particles and precipitated solids. At pH values greater than ten (not shown in Fig. 4.3, but see Fig. 7.1) iron(III) becomes soluble as the $[Fe(OH)_4]^-$ complex assumes predominance.

4.3.5.2 Speciation of iron(III) and sulfur(VI) in the iron(III)-sulfur(VI) system

If sulfur(VI) is present in an aqueous iron(III) solution the complexes, $[FeSO_4]^+$, $[FeHSO_4]^{2+}$ and $[Fe(SO_4)_2]^-$, (all the possible coordinated water molecules being omitted) contribute to the speciation of the iron(III) in addition to the hydrolysis products of Fe(1,0). The speciation of the sulfur(VI) is concerned with those ions together with the sulfate and hydrogen sulfate ions. The four equilibria to be considered are represented by equation (4.7) together with the following equations:

$$Fe(1,0) + SO_4^{2-} \rightleftharpoons [FeSO_4]^+ \qquad (4.48)$$

$$Fe(1,0) + HSO_4^- \rightleftharpoons [FeHSO_4]^{2+} \qquad (4.49)$$

$$[FeSO_4]^+ + SO_4^{2-} \rightleftharpoons [Fe(SO_4)_2]^- \qquad (4.50)$$

The equilibrium constants are defined by equation (4.8) and the equations:

$$K_{1S} = [Fe(1,S)]/[Fe(1,0)][SO_4^{2-}] \qquad (4.51)$$

$$K_{HS} = [Fe(1,HS)]/[Fe(1,0)][HSO_4^-] \qquad (4.52)$$

$$K_{2S} = [Fe(1,2S)]/[Fe(1,S)][SO_4^{2-}] \qquad (4.53)$$

where Fe(1,S), Fe(1,HS) and Fe(1,2S) represent the complexes $[FeSO_4]^+$, $[FeHSO_4]^{2+}$ and $[Fe(SO_4)_2]^-$ respectively. There are two extra equations which are required in order to solve the speciation problem. These are the equations which define the total concentrations of iron(III), T(Fe), and sulfur(VI), T(S), respectively:

$$T(Fe) = [Fe(1,0)] + [Fe(1,1)] + [Fe(1,2)] + [Fe(1,3)]$$
$$+ [Fe(1,4)] + 2[Fe(2,2)] + [Fe(1,S)] + [Fe(1,HS)] + [Fe(1,2S)] \qquad (4.54)$$

$$T(S) = [S] + [HS] + [Fe(1,S)] + [Fe(1,HS)] + 2[Fe(1,2S)] \qquad (4.55)$$

where S and HS are used to represent the sulfate and hydrogen sulfate ions respectively. There are twelve unknowns and eleven equations (numbers (4.8), (4.33)-(4.37) and (4.51)-(4.55)) which, with knowledge of the total iron(III) and sulfur(VI) concentrations, allow the speciation of the two oxidation states to be

calculated for any value of the hydrogen ion concentration.
A combination of equations (4.8) and (4.51)-(4.53), allows equation (4.55) to be expressed in terms of the sulfate ion concentration, represented by [S]:

$$T(S) = [S] + [H^+][S]/K_{S2} + K_{HS}[Fe(1,0)][H^+][S]/K_{S2}$$
$$+ K_{1S}[Fe(1,0)][S] + 2K_{1S}K_{2S}[Fe(1,0)][S]^2 \quad (4.56)$$

Equation (4.56) is a quadratic in [S] and, replacing [Fe(1,0)] by F as before, can be rearranged to give:

$$2K_{1S}K_{2S}F[S]^2 + \{1 + [H^+]/K_{S2} + K_{HS}F[H^+]/K_{S2} + K_{1S}F\}[S] - T(S) = 0 \quad (4.57)$$

The numerical solution of equation (4.57) is not possible because F is not known. Its algebraic solution may be used to represent [S] in the iron(III) equation derived below. A combination of equations (4.33)-(4.37) and (4.51)-(4.53), allows equation (4.54) to be expressed as:

$$T(Fe) = F + K_1F/[H^+] + \beta_2F/[H^+]^2 + \beta_3F/[H^+]^3 + \beta_4F/[H^+]^4$$
$$+ 2K_1^2K_{22}F^2/[H^+]^2 + K_{1S}F[S] + K_{HS}F[H^+][S]/K_{S2} + K_{1S}K_{2S}F[S]^2 \quad (4.58)$$

The expression for [S] in terms of F and T(S) from equation (4.55) may be used in equation (4.58) which produces an equation in which F is the only unknown providing that the values of [H$^+$], T(S) and T(Fe) are specified. Such an equation can be solved iteratively to give the value of F (the free iron(III) concentration, [Fe(1,0)]) from which the speciation fractions can be calculated.

The iron(III) speciation fractions are given by equations (4.42)-(4.47) plus the following equations:

$$\alpha_{1S} = \alpha_{10}K_{1S}[S] \quad (4.59)$$

$$\alpha_{HS} = \alpha_{10}K_{HS}[H^+][S]/K_{S2} \quad (4.60)$$

$$\alpha_{2S} = \alpha_{10}K_{1S}K_{2S}[S]^2 \quad (4.61)$$

where the α values refer to the fractions of the complexes, Fe(1,S), Fe(1,HS) and Fe(1,2S), respectively.

Typical iron(III) and sulfur(VI) concentrations found in bacterial oxidation systems (e.g. in the bacterial oxidation of pyrite, pyrrhotite or chalcopyrite) are 0.3 M and 0.45 M respectively. The values of the equilibrium constants for the iron(III)-sulfur(VI) complexes are those calculated from data given by Robins (personal communication, 1992). Plots of the α values for the major iron(III) components are shown for these conditions in Fig. 4.4.

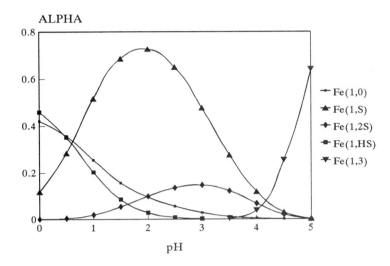

Fig. 4.4 Speciation of the major iron(III) components in a solution of 0.3 M iron(III) and 0.45 M sulfur(VI)

The predominant species in the pH range 1-2 is the sulfatoiron(III) complex, Fe(1,S), with smaller contributions from the unhydrolyzed ion, Fe(1,0), and the hydrogensulfatoiron(III), Fe(1,HS), and bis-sulfatoiron(III), Fe(1,2S), complexes. The affinity between iron(III) and either sulfate or hydrogen sulfate ions depresses the alpha values of the intermediate hydrolysis products of Fe(1,0) so that they are of less importance. As the pH increases beyond a value of around three, the Fe(1,3) complex assumes greater importance and eventually the value of α_{13} approaches unity corresponding with the expected total precipitation of the iron(III) from solution. Under certain conditions compounds known as jarosites may be precipitated. These have the general formula, $MFe_3(SO_4)_2(OH)_6$, where M represents either Na, K or H_3O. The Fe(1,S), Fe(1,HS) and Fe(1,2S) complexes represent precursors of jarosites. Jarosites (discussed further in Chapter 7) are unstable with respect to hydrolysis giving iron(III) oxohydroxide according to the equation:

$$MFe_3(SO_4)_2(OH)_6 + 3OH^- \rightarrow M^+ + 3FeOOH + 2SO_4^{2-} + 3H_2O \qquad (4.62)$$

(an extra OH^- is required in the case of hydronium jarosite to neutralize the H_3O^+ cation resulting in an extra mole of product water). The hydrolysis of the jarosite is a very slow process so its production in bacterial oxidation processes could be avoided by strict control of the pH value of the system.

4.3.5.3 Speciation of iron(III), sulfur(VI) and arsenic(V) in the iron(III)-sulfur(VI)-arsenic(V) system

When arsenic(V) is present in an acidic iron(III) system there are three complexes which have been identified (Glastras, 1988, and Khoe and Robins, 1988) as contributing to the speciation. These are $[FeH_2AsO_4]^+$, $[FeHAsO_4]^{2+}$ and $[FeAsO_4]$ (referred to in the following equations by the respective abbreviations, Fe(1,As2H), Fe(1,AsH), and Fe(1,As)). The neutral $[FeAsO_4]$ may be regarded as an aqueous complex which is the precursor of solid iron(III) arsenate(V) which could be precipitated under appropriate conditions. At pH values greater than five the participation of the bis-arsenatoiron(III) complex, $[Fe(AsO_4)_2]^{3-}$ (abbreviated to Fe(1,2As)), is important (Robins, 1990) in that it leads to the eventual solubilization of iron(III) and arsenic(V). The conditional formation equilibria may be written as:

$$Fe(1,0) + H_2AsO_4^- \rightleftharpoons Fe(1,As2H) \qquad (4.63)$$

$$Fe(1,0) + HAsO_4^{2-} \rightleftharpoons Fe(1,AsH) \qquad (4.64)$$

$$Fe(1,0) + AsO_4^{3-} \rightleftharpoons Fe(1,As) \qquad (4.65)$$

$$Fe(1,0) + 2AsO_4^{3-} \rightleftharpoons Fe(1,2As) \qquad (4.66)$$

The equilibrium constants are defined by the equations:

$$K_{As2H} = [Fe(1,As2H)]/[Fe(1,0)][H_2AsO_4^-] \qquad (4.67)$$

$$K_{AsH} = [Fe(1,AsH)]/[Fe(1,0)][HAsO_4^{2-}] \qquad (4.68)$$

$$K_{As} = [Fe(1,As)]/[Fe(1,0)][AsO_4^{3-}] \qquad (4.69)$$

$$K_{2As} = [Fe(1,2As)]/[Fe(1,0)][AsO_4^{3-}]^2 \qquad (4.70)$$

The path to the solution of the speciation in this complex system is (i) the calculation of the sulfate ion concentration, (ii) the calculation of the concentration of arsenic(V) acid, H_3AsO_4, and (iii) the calculation of the concentration of [Fe(1,0)]. All the alpha values may then be calculated from the value of [Fe(1,0)] (F) and the values of the concentrations of arsenic(V) acid ([As]) and sulfate ion ([S]) where appropriate.

(i) The concentration of sulfate ion.

The presence of arsenic(V) in the system does not change the equations which lead to the calculation of the concentration of sulfate ion. Equation (4.57) remains unchanged with its algebraic solution being used in the iron(III) equation derived

Sec. 4.3] Speciation 89

below. The existence of arsenic(V) complexes of iron(III) introduces a competition for iron(III) so that the alpha values for the sulfate complexes are lower in the presence of arsenic(V) than in its absence.

(ii) The concentration of arsenic(V) acid.

The total concentration of arsenic(V) in the system, T(As), is given by:

$$T(As) = [H_3AsO_4] + [H_2AsO_4^-] + [HAsO_4^{2-}] + [AsO_4^{3-}]$$
$$+ [Fe(1,As2H)] + [Fe(1,AsH)] + [Fe(1,As)] + [Fe(1,2As)] \quad (4.71)$$

Equations (4.13)-(4.15) from the arsenic(V) acid speciation section and equations (4.67)-(4.70) allow equation (4.71) to be reduced to terms which only include $[H_3AsO_4]$ (abbreviated to [As] for the calculations) as far as arsenic(V) is concerned. With rearrangement the equation may be written as:

$$K_{2As}(K_{As1}K_{As2}K_{As3})^2 F[As]^2/[H^+]^6 + [As]\{1 + K_{As1}(1 + K_{As2H}F)/[H^+])$$
$$+ K_{As1}K_{As2}(1 + K_{AsH}F)/[H]^2$$
$$+ K_{As1}K_{As2}K_{As3}(1 + K_{As}F)/[H^+]^3\} - T(As) = 0 \quad (4.72)$$

The algebraic solution of the quadratic equation (4.72) is used to give the value of [As] in the iron(III) equation derived below.

(iii) The concentration of Fe(1,0).

The equation representing the total iron(III) concentration and its components is:

$$T(Fe) = [Fe(1,0)] + [Fe(1,1)] + [Fe(1,2)] + 2[Fe(2,2)] + [Fe(1,3)]$$
$$+ [Fe(1,4)] + [Fe(1,S)] + [Fe(1,SH)] + [Fe(1,2S)] + [Fe(1,As2H)]$$
$$+ [Fe(1,AsH)] + [Fe(1,As)] + [Fe(1,2As)] \quad (4.73)$$

Equation (4.73) may be reduced to one containing only [Fe(1,0)] (abbreviated as before to F) by incorporating equations (4.16)-(4.18) (with added As subscripts to identify the dissociation constants of arsenic(V) acid), (4.33)-(4.37), (4.51)-(4.53) and (4.67)-(4.70). The final equation is:

$$2K_1^2 K_{22} F^2/[H^+]^2 + F\{1 + K_1/[H^+] + \beta_2/[H^+]^2 + \beta_3/[H^+]^3$$

$$+ \beta_4/[H^+]^4 + [S](K_{1S} + K_{HS}[H^+]/K_{S2}) + K_{1S}K_{2S}[S]^2$$

$$+ K_{As1}[As]/[H^+](K_{As2H} + K_{AsH}K_{2As}/[H^+] + K_{As}K_{As2}K_{As3}/[H^+]^2)\}$$

$$+ K_{2As}(K_{As1}K_{As2}K_{As3})^2[As]/[H^+]^6 - T(Fe) = 0 \qquad (4.74)$$

Equation (4.74) can only be solved iteratively after introducing expressions for the values of [S] and [As] given by the algebraic solutions of equations (4.57) and (4.72) respectively.

For given values of T(S), T(As) and T(Fe), solutions of equation (4.74) at various pH values gives the concentrations of free sulfate ion, [S], neutral arsenic(V) acid, [As], and free unhydrolyzed iron(III), F, ([Fe(1,0)]). Those concentrations may then be used to compute the values of the fractions of all the species occurring in the system. The α values for the iron(III) species not containing arsenic are given by equations (4.42)-(4.46) and (4.59)-(4.61) with the substitution of the value of F, and that of [S] where appropriate, from the solutions of equations (4.57) and (4.74). The α values for the iron(III) complexes containing arsenic are given by the equations:

$$\alpha_{As2H} = \alpha_{10} K_{As2H} K_{As1}[As]/[H^+] \qquad (4.75)$$

$$\alpha_{AsH} = \alpha_{10} K_{AsH} K_{As1} K_{As2}[As]/[H^+]^2 \qquad (4.76)$$

$$\alpha_{As} = \alpha_{10} K_{As} K_{As1} K_{As2} K_{As3}[As]/[H^+]^3 \qquad (4.77)$$

$$\alpha_{2As} = \alpha_{10} K_{2As}(K_{As1} K_{As2} K_{As3})^2[As]^2/[H^+]^6 \qquad (4.78)$$

The alpha values for the major iron(III) components of a solution which contains 0.3 M iron(III), 0.45 M sulfur(VI) and 0.2 M arsenic(V) (typical of the solutions produced in the bacterial oxidation of arsenopyrite/pyrite mixtures) are plotted as a function of pH for pH values from zero to two in Fig. 4.5. The onset of the formation of the neutral Fe(1,As) complex is also indicated. The values of the formation constants of the iron(III)-arsenic(V) complexes were calculated from thermodynamic data supplied by Robins (1992). There are four complexes which are important; Fe(1,AsH), Fe(1,S), Fe(1,As2H) and Fe(1,HS). They account for around 90% of the iron(III) with any other complexes being of little importance. As the pH value increases the production of the neutral Fe(1,As) complex begins to be significant which is consistent with the onset of precipitation of solids which is observed to be below a pH value of around two. The nature of the solids which are precipitated from Fe-S(VI)-As(V) solutions as the pH is raised is discussed in Chapter 7.

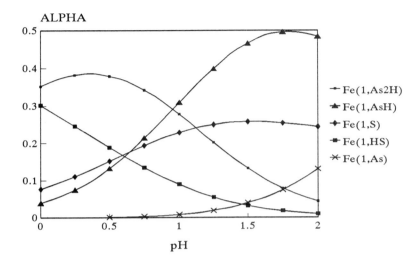

Fig. 4.5 Speciation of the major iron(III) components in a solution of 0.3 M iron(III), 0.45 M sulfur(VI) and 0.2 M arsenic(V)

4.3.6 Other solubilized metals

Taking the metals considered in Chapter 2 as examples for this section, the species present (separately or as various mixtures) in the solution, from the bacterial oxidation of their appropriate minerals, would be Ag^+, Co^{2+}, UO_2^{2+}, Ni^{2+}, MoO_2^{2+}, Cu^{2+}, Zn^{2+} and Pb^{2+}. The majority of any silver (from silver sulfide), lead, tin or antimony would be present as sparingly soluble solids. In addition there would be the ions of the nutrients present as well as iron(III) (from the oxidation of pyrite/arsenopyrite), sulfur(VI) (from sulfide oxidation) and arsenic(V) (from the oxidation of any arsenic compounds). If lime or limestone was used to adjust the pH of the system the Ca^{2+} ion would be present in the system.

The speciation of any aqueous ion produced in the bacterial oxidations, except that of iron(III), is not of great importance. The complexation of metal ions by the organic effluent products of bacterial metabolism is discussed in Chapter 3 with regard to its possibility of reducing metal toxicities. None of the hydrated Ag^+, Co^{2+}, UO_2^{2+}, Ni^{2+}, Cu^{2+}, Zn^{2+} and Pb^{2+} ions is hydrolyzed at the pH values appropriate to their production in bacterial oxidation reactions, their lowest pK values for hydrolysis all being greater than two (Perrin, 1969e). The speciation of antimony, molybdenum and tin is dealt with in sections 2.2.9, 2.2.6 and 2.2.7 respectively. The interaction of iron(III) with molybdenum(VI) is referred to in section 2.2.6. The interactions of some importance are those between the product ions and sulfate and/or arsenic(V). Silver(I) ion, calcium(II) and lead(II) ions all form insoluble sulfates. All the ions in the (II) state form insoluble arsenates with the formulae, $M_3(AsO_4)_2$, most of which exist as authenticated minerals (M = Ca, Pb, Zn, Ni, and Cu) (Crecelius et al. 1986). The presence of arsenic in such a system could possibly complicate the eventual recovery of the metal values. Further discussion of the speciation of the elements dealt with in this section is contained in Chapter 7 (section 7.3).

4.4 CHEMICAL EQUATIONS

With the above considerations it is clearly not possible to write single chemical equations to express all the aspects of the overall processes for the aqueous oxidations of pyrite and arsenopyrite. The simplest equations which express the oxidations and do not take hydrolysis and complexation into account are:

$$4FeS_2 + 15O_2 + 2H_2O + 4H^+ \longrightarrow 4Fe^{3+} + 8HSO_4^- \qquad (4.79)$$

$$2FeAsS + 7O_2 + 2H_2O + 4H^+ \longrightarrow 2Fe^{3+} + 2H_3AsO_4 + 2HSO_4^- \qquad (4.80)$$

In both cases there is an increase in the number of acidic protons which leads to a general decrease in the pH of the solution as the reactions proceed. Exception to that statement can occur if the gangue material possesses ion exchange properties and acts as a proton sink. Basic and carbonate gangues would also counteract the tendency to lower pH values by the oxidation reactions.

In the literature there are variations upon equations (4.79) and (4.80) which are less precise. The products of equation (4.79), minus four protons, may be re-arranged to give, $2Fe_2(SO_4)_3 + 2H_2SO_4$, both representing substances which dissociate in aqueous solution. The products of equation (4.80), again minus the four protons, are sometimes re-arranged to give, $2FeAsO_4 + 2H_2SO_4$, the first component being a solid which is only stable at pH values greater than 2.2. The nature of the precipitated material when the pH is raised to a value greater than two is discussed in Chapter 7. Equations (4.79) and (4.80) represent overall changes and have no bearing on the mechanisms of the bacterial oxidation processes.

There are observable intermediates in the reaction mixtures which supply evidence for the bacterial oxidation mechanisms. In a typical bacterial oxidation of a pyrite/arsenopyrite mixture there are small concentrations of iron(II) (present as the $[Fe(H_2O)_6]^{2+}$ ion) and arsenic(III) (present as the neutral arsenic(III) acid, H_3AsO_3) (Barrett et al. 1989). Small amounts of elementary sulfur are detectable in bacterial oxidation reactions (Barrett et al. 1988), but no other intermediate species have been detected.

4.5 ELECTRODE POTENTIALS

In bacterial oxidation systems there are two oxidants to be considered. These are oxygen, O_2, and iron(III) in its various complexed forms. The oxidizing power of an oxidant is expressed by the value of its standard reduction potential, $E^{\ominus}$. High values of the standard reduction potential indicate that the oxidized form of the couple (of oxidation states) is a good oxidizing agent. The oxidizing power of an oxidant under non-standard conditions is expressed by the appropriate Nernst equation. In all cases the units of the terms in the Nernst equations are volts.

4.5.1 The oxidizing power of oxygen

The reduction of the oxygen molecule to liquid water may be written as:

$$O_2(g) + 4H^+(aq) + 4e^- \rightarrow 2H_2O(l) \tag{4.81}$$

The standard reduction potential for this half-reaction is 1.23 V and the Nernst equation, which expresses the variation of its potential, E, with the hydrogen ion activity is:

$$E = E^{\ominus} - (RT/4F)\ln(1/(p_{O_2} a_{H^+}^4)) \tag{4.82}$$

For conditions where the partial pressure of oxygen (p_{O_2}) is unity (the standard state) the substitution of the pH value as given by equation (4.1) gives (at 298K):

$$E = E^{\ominus} - (2.303RT/F).pH = 1.23 - 0.0592.pH \tag{4.83}$$

The scale of values of $E^{\ominus}$ (based upon the reduction of the aqueous proton to hydrogen as the reference zero) varies from 3.06 V (for the reduction of fluorine to hydrogen fluoride) to around -3 V (for the reduction of the Group 1 uni-positive aqueous ions, Li^+ to Cs^+, to their metallic state). Oxygen has a moderately strong oxidizing capacity according to such a scale. In the metabolism of bacteria, oxygen operates in the cytoplasmic membrane where the local pH value is considered to be around seven. At this value of pH the value of the reduction potential, as given by equation (4.83), is 0.816 V. It is this value which should be used in comparisons of the relative oxidizing capacities of oxygen and other possibly involved oxidants.

4.5.2 The oxidizing power of iron(III)

The half-reaction for the reduction of Fe^{3+} to Fe^{2+} has a standard potential of 0.77 V, which indicates that the iron(III) tripositive cation is a less powerful oxidizing agent than oxygen. The Nernst equation for the iron(III)-iron(II) reduction shows no dependence upon pH since the half-reaction does not involve aquated protons. The equation ($E = 0.77$ V) is, however, of very limited application because of (i) the hydrolysis of iron(III) at low values of pH and of iron(II) at pH values greater than 6.8 and (ii) the existence of iron(III)-sulfur(VI) and iron(III)-arsenate(V) complexes in solutions of practical importance.

4.5.2.1 The effects of sulfur(VI) and arsenic(V) complexation upon the oxidizing power of iron(III)

In the pH range from zero to two the important forms of iron(III) in practical systems are its sulfato- and arsenato-complexes. The effect of complexation of the iron(III) upon the standard reduction potential of the iron(III)/iron(II) couple may be calculated from known thermodynamic data. In the presence of sulfate ion the complex $[FeSO_4]^+$ is one of the ions formed, the half-reaction for its reduction to iron(II) being:

$$[FeSO_4]^+ + H^+ + e^- \rightarrow Fe^{2+} + HSO_4^- \tag{4.84}$$

The full Nernst equation for this half-reaction is:

$$E = 0.684 - \frac{RT}{F} \ln \frac{a_{Fe^{2+}} \cdot a_{HSO_4^-}}{a_{FeSO_4^+} \cdot a_{H^+}} \tag{4.85}$$

At a temperature of 298 K the equation may be written as:

$$E = 0.684 - 0.0592 \cdot pH - 0.0592 \cdot \log \frac{a_{Fe^{2+}} \cdot a_{HSO_4^-}}{a_{FeSO_4^+}} \tag{4.86}$$

If the activities of all the ions, except for that of the hydrogen ion, are assumed to be equal to unity (their standard states), the equation simplifies to:

$$E = 0.684 - 0.0592 \cdot pH \tag{4.87}$$

Calculations from appropriate data for the half-reactions for the reduction of other iron(III) complexes shown below allow their Nernst equations to be derived. The equations given assume that the activities of all the ions concerned, with the exception of that of the hydrogen ion, are unity, and that the temperature is 298 K.

$$[FeHSO_4]^{2+} + e^- \rightarrow Fe^{2+} + HSO_4^- \tag{4.88}$$

$$E_{4.88} = 0.709 \tag{4.89}$$

Equation (4.89) does not contain a term involving the pH value since the half-reaction does not contain any free aquated protons.

$$[Fe(SO_4)_2]^- + 2H^+ + e^- \rightarrow Fe^{2+} + 2HSO_4^- \tag{4.90}$$

$$E_{4.90} = 0.767 - 0.1182 \cdot \text{pH} \qquad (4.91)$$

$$[\text{FeH}_2\text{AsO}_4]^{2+} + \text{H}^+ + e^- \rightarrow \text{Fe}^{2+} + \text{H}_3\text{AsO}_4 \qquad (4.92)$$

$$E_{4.92} = 0.598 - 0.0592 \cdot \text{pH} \qquad (4.93)$$

$$[\text{FeHAsO}_4]^+ + 2\text{H}^+ + e^- \rightarrow \text{Fe}^{2+} + \text{H}_3\text{AsO}_4 \qquad (4.94)$$

$$E_{4.94} = 0.654 - 0.1182 \cdot \text{pH} \qquad (4.95)$$

The iron(III) complex with the lowest potential over the pH range from zero to two is $[\text{FeH}_2\text{AsO}_4]^{2+}$ (under the standard conditions defined above) and is, in consequence, the poorest oxidant over this pH range. Such an observation is consistent with the higher formation constants of the arsenato-complexes as compared to the sulfato-complexes. The argument which is used in the next section is that if the dihydrogenarsenato(V) iron(III) complex has the potential to oxidize a mineral to a particular level then so will the other iron(III) complexes.

4.6 THERMODYNAMICS OF MINERAL OXIDATION

A possibly very important role for iron(III) in bacterially catalysed oxidation reactions is its action as an oxidizing agent. It may be considered as an oxidant for the refractory minerals and also for any arsenic(III) which is produced as an intermediate in the oxidation of any arsenic content to arsenic(V). Pyrite, arsenopyrite and chalcopyrite are taken as examples of mineral oxidations after which the oxidation of arsenic(III) is considered. The Nernst equations which are quoted in the following sections are those for which the activities of all ions except that of the proton are assumed to be unity.

4.6.1 The oxidation of pyrite

Pyrite, FeS_2, may be considered formally to contain iron(II) and sulfur(-1) and may be oxidized in aqueous solution to either (i) sulfur(0) with the release of iron(II) into the solution, (ii) sulfur(VI) with release of iron(II), or (iii) iron(III) and sulfur(VI). The iron(III) complex chosen to represent case (iii) is the sulfatoiron(III) ion, $[\text{FeSO}_4]^+$, because it has the lowest standard reduction potential with respect to its reduction to iron(II). The appropriate half-reactions with their derived Nernst equations (with the restrictions as in the above cases) are now given.

$$\text{Fe}^{2+} + 2\text{S} + 2e^- \rightarrow \text{FeS}_2 \qquad (4.96)$$

$$E_{4.96} = 0.424 \qquad (4.97)$$

$$\text{Fe}^{2+} + 2\text{HSO}_4^- + 14\text{H}^+ + 14e^- \rightarrow \text{FeS}_2 + 8\text{H}_2\text{O} \qquad (4.98)$$

$$E_{4.98} = 0.351 - 0.0592 \cdot pH \qquad (4.99)$$

$$FeSO_4^+ + HSO_4^- + 15H^+ + 15e^- \rightarrow FeS_2 + 8H_2O \qquad (4.100)$$

$$E_{4.100} = 0.373 - 0.0592 \cdot pH \qquad (4.101)$$

The standard reduction potentials (pH = 0) are summarized in the diagram shown in Fig. 4.6.

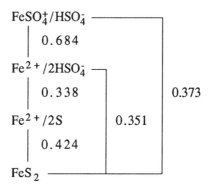

Fig.4.6 Standard reduction potentials (volts) appropriate to the oxidation of pyrite

The oxidation of pyrite to sulfur(VI), with the solubilization of iron(II), is feasible with any of the iron(III)/sulfate/arsenate complexes, the final stage of oxidation of iron(II) to iron(III) requiring oxygen to be the oxidant. Although the iron(III) complexes appear to have the potential to oxidize pyrite through to the iron(III) stage the iron would be produced in the (II) state by the reduction of the iron(III) complex.

4.6.2 The oxidation of chalcopyrite

Chalcopyrite, $CuFeS_2$, may be regarded formally to contain copper(II), iron(II) and sulfur(-2). It may be oxidized to either (i) sulfur(0) with the release of copper(II) and iron(II) into solution, (ii) sulfur(VI) with the release of the copper(II) and iron(II) into the solution, or (iii) sulfur(VI) and iron(III) with release of the copper(II) into the solution. As in the case of pyrite oxidation (section 4.6.1) the form of iron(III) produced in case (iii) is considered to be the sulfato-complex. Calculations from thermodynamic data allow the derivation of the Nernst equations for the selected half-reactions below. They are quoted for the cases where all ions except the hydrogen ion are at unit activities and for a temperature of 298 K.

$$Cu^{2+} + Fe^{2+} + 2S + 4e^- \rightarrow CuFeS_2 \quad (4.102)$$

$$E_{4.102} = 0.397 \quad (4.103)$$

$$Cu^{2+} + Fe^{2+} + 2HSO_4^- + 14H^+ + 16e^- \rightarrow CuFeS_2 + 8H_2O \quad (4.104)$$

$$E_{4.104} = 0.353 - 0.0517 \cdot pH \quad (4.105)$$

$$Cu^{2+} + FeSO_4^+ + HSO_4^- + 15H^+ + 17e^- \rightarrow CuFeS_2 + 8H_2O \quad (4.106)$$

$$E_{4.106} = 0.374 - 0.0487 \cdot pH \quad (4.107)$$

The diagram shown in Fig. 4.7 summarizes the standard reduction potentials (pH = 0) for the oxidation of chalcopyrite.

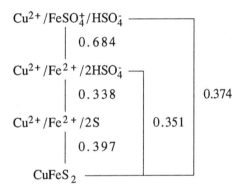

Fig.4.7 Standard reduction potentials (volts) appropriate to the oxidation of chalcopyrite

As in the case of pyrite oxidation, the iron(III) complexes have the potential to cause the oxidation of chalcopyrite to the iron(II)/sulfur(VI) stage, oxygen being required for the final oxidation to iron(III).

4.6.3 The oxidation of arsenopyrite

Calculated values of the standard reduction potentials and the corresponding Nernst equations (with the activities of all ions except that of the hydrogen ion being unity), at 298 K, for the reduction of (i) iron(II), arsenic(III) acid and sulfur(0), (ii) iron(II), arsenic(III) acid and sulfur(VI), (iii) iron(II) arsenic(V) acid and sulfur(VI), and (iv) dihydrogenarsenato(V)iron(III) and sulfur(VI) to solid arsenopyrite and liquid water are given below, with the appropriate half-reactions.

$$3H^+ + Fe^{2+} + H_3AsO_3 + S + 5e^- \rightarrow FeAsS + 2H_2O \qquad (4.108)$$

$$E_{4.108} = 0.077 - 0.0355.pH \qquad (4.109)$$

$$10H^+ + Fe^{2+} + H_3AsO_3 + HSO_4^- + 11e^- \rightarrow FeAsS + 6H_2O \qquad (4.110)$$

$$E_{4.110} = 0.219 - 0.0537.pH \qquad (4.111)$$

$$12H^+ + Fe^{2+} + H_3AsO_4 + HSO_4^- + 13e^- \rightarrow FeAsS + 8H_2O \qquad (4.112)$$

$$E_{4.112} = 0.274 - 0.0546.pH \qquad (4.113)$$

$$13H^+ + FeH_2AsO_4^{2+} + HSO_4^- + 14e^- \rightarrow FeAsS + 8H_2O \qquad (4.114)$$

$$E_{4.114} = 0.296 - 0.0549.pH \qquad (4.115)$$

The standard reduction potentials (pH = 0) are summarized in the diagram shown in Fig. 4.8. The iron(III) complexes have the potential to oxidize arsenopyrite to the iron(II)/sulfur(VI) stage. The oxidation of arsenic(III) to arsenic(V) is dealt with in the next section. Oxygen is required for the final stage of oxidation of the iron(II) to iron(III).

4.6.4 The oxidation of arsenic(III) to arsenic(V)

The standard reduction potential for the half-reaction:

$$H_3AsO_4 + 2H^+ + 2e^- \rightarrow H_3AsO_3 + H_2O \qquad (4.116)$$

is generally accepted to be 0.56 V (Latimer, 1955b), although calculations from recent data (which are consistent with those given by Latimer) give a value of 0.577 V. The latter value is used to preserve the consistency of the calculations reported in this chapter. The Nernst slope at 298 K is 0.0592 until the first dissociation of arsenic(V) acid becomes important (pH greater than 2.2). When arsenic is present in a bacterial oxidation system the iron(III) in solution is present mainly as arsenic(V) complexes. At a pH of 1.5 (that corresponding to most practical applications) the predominant arsenic(V) complex of iron(III) is the hydrogenarsenato(V)iron(III) complex, $[FeHAsO_4]^+$, as is shown in Fig. 4.5. At that pH value the reduction potential of the complex, with respect to iron(II) and hydrogenarsenate(V) ion, is given by equation (4.95) as 0.477 V and that for the reduction of arsenic(V) to arsenic(III) by equation (4.117) as 0.488 V implying that the oxidation of arsenic(III) by iron(III) is unlikely to be a feasible process.

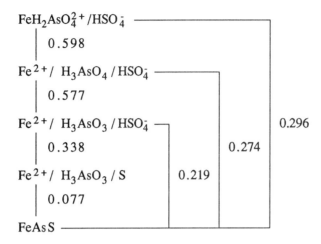

Fig. 4.8 Standard reduction potentials (volts) appropriate to the oxidation of arsenopyrite

The Nernst equation for the restricted pH range is:

$$E_{4.116} = 0.577 - 0.0592.\text{pH} \qquad (4.117)$$

At equilibrium (if that state were to be established) there would be a considerable fraction of unoxidized arsenic(III) in the system. The reduction potential of the dihydrogenarsenato(V)iron(III) complex, at a pH value of 1.5, is given by equation (4.93) as 0.509 V which is just sufficient to favour the feasibility of the oxidation of arsenic(III) to arsenic(V). There is a more decisive difference between the reduction potential of the sulfatoiron(III) complex (equation (4.87) gives a value of 0.595 V) and that of arsenic(V) at a pH value of 1.5. The hydrogensulfatoiron(III) complex also has the potential to oxidize arsenic(III) to arsenic(V). It should be pointed out that none of the differences in standard potentials is very large and that what happens in practice will depend upon local conditions in the system. None of the activities will be unity and it may very well be that the oxidation of arsenic(III) is dependent upon some interaction with a surface (either bacterial or mineral or both). Oxygen does have the necessary high potential to make the arsenic(III) to arsenic(V) oxidation feasible, but the reaction is known to be very slow. This is presumably because the oxidation of arsenic(III) is a two-electron change dependent upon the transfer of an oxygen atom, the reduction of oxygen being a four-electron process. Such uncomplementarity is well known to cause reactions to be slow. Further discussion of the oxidation of arsenic(III) is contained in section 5.3.3.2.

4.6.5 Thermodynamic conclusions

The thermodynamic conclusions which may be drawn from the above calculations are that aqueous iron(III), in its various complex forms, has the potential to cause the oxidations of (i) pyrite to sulfur(VI) with the solubilization of iron(II), (ii) chalcopyrite to sulfur(VI) with the solubilization of Cu(II) and Fe(II), (iii) arsenopyrite to arsenic(V) and sulfur(VI) with the solubilization of Fe(II). The latter conclusion (with regard to the production of arsenic(V)) is made on the basis of small differences between the standard reduction potentials concerned and is really dependent upon the observation that oxygen oxidizes arsenic(III) to arsenic(V) very slowly. Some of the forms of iron(III) in the bacterial oxidation system (sulfato-complexes) have the potential to cause the oxidation of arsenic(III) to arsenic(V) but only oxygen has the potential to cause the further oxidation of aqueous iron(II) to iron(III). Such conclusions about reaction feasibilities do not necessarily imply any mechanistic involvement of the reactions considered.

It should be emphasized at this stage that the standard potentials discussed have been *calculated* from thermodynamic quantities (standard Gibbs energies of formation and equilibrium constants at 298 K). At no stage have electrode potentials been actually *measured*. It is most unlikely that any electrode measurements have any meaning. For them to be useful it is essential that the reactions studied are *reversible* in the thermodynamic sense of that term. No kinetic conclusions should be drawn from the magnitude of an overall standard potential however positive the value may be although violations of this rule are very common in the literature.

4.7 CHEMICAL OXIDATIONS

It is useful to compare the rates of bacterial oxidation of minerals with those of chemical processes. The rates of oxidation of pyrite, arsenopyrite and chalcopyrite by oxygen at ambient temperatures are immeasurably slow. It is important to consider the effects of acids and acidic iron(III) solutions on these minerals to emphasize the necessity for catalytic bacteria to be present for efficient oxidation.

4.7.1 Chemical dissolution of chalcopyrite

In the absence of oxygen (experiments carried out by bubbling nitrogen gas through the reaction mixture) acid attack on chalcopyrite (Ichikuni, 1960) causes the solubilization of iron(II), the evolution of hydrogen sulfide, H_2S, and leaves the copper in the solid state as CuS:

$$FeCuS_2 + 2H^+ \rightarrow Fe^{2+} + CuS + H_2S \qquad (4.118)$$

The reaction is very slow even at 80-85°C under which conditions the use of 0.1 M HCl caused less than 0.5% reaction. When the experiment was carried out in air the solubilization of the iron(II) rose to three percent and the copper was also solubilized. Some oxidation of the sulfide sulfur to sulfate occurred.

Iron(III), in the presence of air, does increase the solubilization rate of chalcopyrite (Habashi, 1978) but the reaction rate is still too slow to be exploited economically. The effect of a 0.36 M solution of iron(III) sulfate at 35°C upon -325 mesh chalcopyrite was to solubilize thirteen percent of its copper content in ten days. In the presence of a suitable bacterial culture the reaction is complete within twelve to twenty four hours.

4.7.2 Chemical dissolution of pyrite and arsenopyrite

Both pyrite and arsenopyrite may be considered as containing iron(II) which would be solubilized (but not oxidized) by the chemical attack by iron(III). Such an attack would oxidize the sulfur to sulfur(VI) and the arsenic to arsenic(III) with concurrent release of the iron(II) into the solution. The chemical attack on pyrite is extremely slow and is accelerated by a factor of the order of 10^6 by suitable bacteria (Silverman and Ehrlich, 1964). The bacteria catalyse the reaction and enable the iron(II) to be oxidized to iron(III). The action of iron(III), in an acid sulfate solution, upon arsenopyrite has been shown (Barrett *et al*, 1990) to follow the equation:

$$FeAsS + 11Fe(III) \rightarrow 12Fe(II) + As(III) + S(VI) \quad (4.119)$$

Under very similar conditions the bacterially catalysed process is at least six times faster than the chemical one and, in addition, causes the oxidation of the iron(II) to iron(III) and arsenic(III) to arsenic(V) (Barrett *et al.* 1989).

4.8 REFERENCES

Atkins, P.W. (1982) *Physical chemistry.* 2nd edn, p.355.
Baes, C.F. & Mesmer, R.E. (1976) *The Hydrolysis of Cations.* John Wiley & Sons. p.235.
Barrett, J., Ewart, D.K., Hughes, M.N., Nobar, A.M., O'Reardon, D.J. & Poole, R.K. (1988) *R & D for the Minerals Industry, Kalgoorlie, 1988*, Western Australian School of Mines, p.275.
Barrett, J., Ewart, D.K., Hughes, M.N., Nobar, A.M. & Poole, R.K. (1989) *Biohydrometallurgy - 89, Jackson Hole, 1989*, Salley, J., McReady R.G.L. & Wichlacz, P.L. (eds), p.49.
Barrett, J., Hughes, M.N. & Russell, A. (1990) *Randol Gold Forum, Squaw Valley, 1990*, p.135.
Crecelius, E.A., Bloom, N.S., Cowan, C.E. & Jenne, E.A. (1986) *Speciation of Selenium and Arsenic in Natural Waters and Sediments, Volume 2: Arsenic Speciation*, EPRI EA-4621, Volume 2, Research Project 2020-2, p.1-3.

Dutrizac, J.E. & Monhemius, A.J. (eds) (1986) *Iron Control in Hydrometallurgy.* Ellis Horwood.
Flynn, C.M. (1984) *Chem. Rev.,* **84**, 31.
Glasstone, S. (1947) *Thermodynamics for Chemists.* Van Nostrand, p.402.
Habashi, F. (1978) *Chalcopyrite: Its Chemistry and Metallurgy.* McGraw-Hill, p.80.
Hall, H.S. & Knight, S.R. (1948) *Higher Algebra.* Macmillan, p.83.
Ichikuni, M. (1960) *Bull. Chem. Soc. Japan* **33**, 1052.
Khoe, G.H., Brown, P.L., Sylva, R.N. & Robins, R.G. (1986) *J. Chem. Soc. Dalton Trans.,* p.1901.
Khoe, G.H. & Robins, R.G. (1988) *J. Chem. Soc. Dalton Trans.,* p.2015.
Latimer, W.L. (1955) *Oxidation Potentials.* Prentice Hall, (a) p.113, (b) p.116.
Perrin, D.D. (1969) *Dissociation Constants of Inorganic Acids and Bases in Aqueous Solution.* IUPAC, Butterworths, (a) p.200, (b) p.147, (c) p.146, (d) p.177 and (e) p.151, 158, 179, 183, 198, 208, 210 and 216.
Robins, R.G. (1990) *EPD Congress '90, TMS, Anaheim, 1990,* Gaskell, D.R.(ed.) p.93.
Robins, R.G. (1992) *personal communication*
Silverman, M.P. & Ehrlich, H.L. (1964) *Advan. Appl. Microbiol.,* **6**, 153.

5

The general mechanism of bacterial oxidation

5.1 INTRODUCTION

There are three areas of discussion which are concerned with the mechanisms of bacterial oxidation processes. The first of these is the breakdown of the overall chemical reaction equation into various stages, each of which is a properly balanced chemical equation and actually contributes to the reaction path. If possible each stage should be identified as being either bacterially catalysed or an abiotic chemical reaction. The second area is the consideration of the detailed mechanism of each of the stages. The third area of discussion is that concerned with the mechanism of the participation, in any of the appropriate stages, of the particular catalytic bacterium. The mechanisms used by the bacteria to catalyse oxidation reactions are discussed in Chapter 3.

Only those aspects of the detailed mechanisms of the overall reactions for the oxidations of pyrite and arsenopyrite which are relevant to the design of bacterial oxidation plant are discussed in this chapter. The elucidation of the main general features of the mechanism of the oxidations leads to the identification of parameters whose measurement allows the control of plant using the process. No treatment of the increasingly large number of publications which deal with mathematical modelling of the kinetics of bacterial oxidation reactions is included. The book by Rossi (1990) contains a review of the literature concerned with attempts to model the process.

5.2 PREVIOUS MECHANISTIC CONCLUSIONS

The literature concerned with the mechanistic aspects of bacterially catalysed oxidation of minerals has been reviewed by Kelly *et al.* (1979).

The oxidation of pyrite, expressed by the overall equation:

$$4FeS_2 + 15O_2 + 2H_2O + 4H^+ \rightarrow 4Fe(III) + 8HSO_4^- \tag{5.1}$$

was considered to occur in four stages (Panin *et al.* 1985) which were written as the equations:

$$2FeS_2 + 7O_2 + 2H_2O \rightarrow 2FeSO_4 + 2H_2SO_4 \qquad (5.2)$$

$$4FeSO_4 + O_2 + 2H_2SO_4 \rightarrow 2Fe_2(SO_4)_3 + 2H_2O \qquad (5.3)$$

$$FeS_2 + 2Fe^{3+} \rightarrow 3Fe^{2+} + 2S \qquad (5.4)$$

$$2S + 3O_2 + 2H_2O \rightarrow 2H_2SO_4 \qquad (5.5)$$

with reactions (5.2), (5.3) and (5.5) being bacterially catalysed. Because equation (5.2) is considered to be bacterially catalysed the reaction is regarded as being an example of **direct** bacterial oxidation. Reaction (5.4) was not regarded to be bacterially catalysed but, because the oxidant, iron(III), is bacterially produced by reaction (5.3), it has been described as **indirect** bacterial oxidation. The term indirect is generally applied to those chemical reactions where the oxidizing agent (usually iron(III)) is generated separately in a bacterially catalysed reaction.

The bacterial oxidation of arsenopyrite, as represented by the overall equation:

$$2FeAsS + 7O_2 + 2H_2O + 4H^+ \rightarrow 2Fe(III) + 2H_3AsO_4 + 2HSO_4^- \qquad (5.6)$$

is reported (Shrestha, 1988) to take place in two stages. The first stage is the bacterially catalysed oxidation of arsenopyrite to iron(II), sulfur(VI) and arsenic(III) as represented by the equation:

$$4FeAsS + 11O_2 + 6H_2O \rightarrow 4FeSO_4 + 4H_3AsO_3 \qquad (5.7)$$

This is followed by the alleged chemical oxidation of the arsenic(III) by the iron(III) produced by reaction (5.3):

$$H_3AsO_3 + 2Fe^{3+} + H_2O \rightarrow H_3AsO_4 + 2Fe^{2+} + 2H^+ \qquad (5.8)$$

Reaction (5.7) would be described as being direct, whilst reaction (5.8) would be called indirect, using the above definitions of those terms.

The above mechanistic conclusions were arrived at before more conclusive experimental evidence was available. They are valuable in that they identify intermediates (these are species which are neither initial reactants nor final products) in both cases. The intermediates are iron(II) and sulfur(0) in the case of pyrite oxidation and iron(II) and arsenic(III) in the case of arsenopyrite oxidation. There are apparently two different sulfur oxidation pathways since no elementary sulfur has been reportedly observed in the oxidation of arsenopyrite.

The following section is a discussion of the key points from various investigations which are relevant to the mechanism of the bacterial oxidation of arsenopyrite and pyrite, most recent work having been carried out with mixtures of these two minerals.

5.3 THE MECHANISM OF BACTERIAL OXIDATION REACTIONS

This section contains an introduction to the principles of kinetics as they are generally applied to the deduction of mechanisms of reactions. These are then used to elucidate a general mechanism of bacterial oxidation reactions.

5.3.1 Introduction to mechanistic principles

The mechanism of any reaction is deduced from the rate law for the process. The rate law is the experimentally observed dependence of the rate of the reaction upon any variables which have been found to affect the rate. These normally might include the concentrations of the reactants and possibly those of the products, the concentrations of any catalyst present and the surface area of any solid participating in the process. The rate law for any system is essentially empirical and can not be deduced from the stoichiometric equation representing the overall process.

The main kinetic characteristic of a substance (ion or molecule) taking part in a reaction is the **order** of the reaction with respect to that substance. The order of a reaction with respect to one of its participant substances is the power to which the concentration of that substance is raised in the rate equation. The order of a reaction with respect to the concentration of a reactant, product or catalyst, is shown symbolically by the equation:

$$\text{Rate} = [\text{reactant, product or catalyst}]^{\text{order}} \qquad (5.9)$$

It is the power to which the concentration is raised for each participating substance, in the equation representing the rate of the process. In some reactions (those with complicated rate laws) it is not possible to state the order of any particular substance. In other cases the order is fractional and sometimes equal to zero (when the reaction rate is independent of the concentration of the substance).

Once the rate law of a reaction has been established some mechanistic deductions may be attempted. The guiding principle is that of **maximum simplicity**. This has its origin in the works of William of Occam (early 14th. century philosopher) and is known as Occam's razor, which he stated in the form 'It is vain to do with more what can be done with fewer', (Russell, 1946) or, 'Don't make things unnecessarily complicated' (Barrett, 1992).

In homogeneous (gas or solution phase) reactions the kinetic processes occur as a result of suitably energetic collisions between the participating molecules. Such collisions are normally **bimolecular**, i.e. they take place between two molecules or ions. The mechanism of the reaction is considered to be the minimum number of single stage processes which are consistent with the rate equation. Heterogeneous processes involving a solution and a surface (mineral or bacterial) are more complex. It is normal to attempt to interpret the kinetics of such systems in terms of the transfer of reactants from the solution phase to the solid phase by collisional processes. The crucial part of the deductive process is the abstraction from the rate law of which molecules (or ions or surfaces or bacteria) participate in the **rate determining step**. The rate determining step is the slowest step in the reaction. Every other step in the reaction mechanism is to be considered as being faster than the one which determines the rate. Sometimes, in the literature of biohydrometallurgy, reference is made to a particular step being **rate limiting**. The proper use of that term is applicable to the case of a reaction whose rate is limited by there being such a low concentration of a particular component of the system that its concentration governs the rate of the overall reaction. The particular component may or may not feature in the rate law and may or may not normally contribute to the rate determining step. There is some misuse of the two terms in the literature.

5.3.2 The mechanism of bacterial oxidation

Bacterial oxidation reactions are far from simple and consist of many rate processes, some of which are relatively simple and well understood, others being complex with their understanding still under various stages of development. The various rate processes which contribute to the bacterial oxidation of minerals fall into the following categories.

5.3.2.1 Dissolution of gases in aqueous solution and their aqueous transport

Both oxygen and carbon dioxide are essential for the process of bacterial oxidation. Oxygen is needed as an oxidant and carbon dioxide supplies the bacteria with their carbon requirements. The rates of transfer of oxygen and carbon dioxide from the gaseous phase to the aqueous phase are well known under a variety of conditions. It is important for bacterial oxidation carried out in agitated tank reactors to have a sufficiently rapid transfer of these gases from the atmosphere by having an appropriate air-blowing facility. Other methods of carrying out the reaction (dump, heap, vat, and *in situ* methods) must rely upon the natural solubility of these gases in the solutions used. In heap and dump leaching the air-water interface may be enlarged by spraying the water onto the top of the material. The rates of diffusion of the dissolved gases to the bacteria (either in the bulk solution or on the mineral surface) may be rate limiting under certain circumstances. This would only be the case for agitated tank reactors if the air-blowing rate and/or the stirring rate fell below some threshold value. Normally none of the processes involving the two gases is slow enough to be rate limiting. It may be that the supply of water containing the dissolved gases would be rate limiting in the cases of *in situ* processes which produce acid mine drainage and vat leaching.

The equation which relates the rate of dissolution of a gas by a solution to its solubility is written as:

$$dC_L/dt = K_L.a.(C_{sat} - C_L) \qquad (5.10)$$

where C_L is the concentration of the gas in the liquid phase, C_{sat} is the concentration of a saturated solution of the gas (its equilibrium solubility), K_L is the mass transfer coefficient and a is the effective area of the gas-liquid interface surface. Neither K_L nor a are easily obtained parameters and normally their product, $K_L.a$, is quoted. A typical oxygen usage rate for the bacterial oxidation of arsenopyrite (fifteen percent (w/v) pulp density) is 5.6 mmol L^{-1} h^{-1}. The saturation solubility of oxygen in water at 35°C is 0.4 mmol L^{-1} and if the bacterial oxidation solution is to have a concentration of half that value the $K_L.a$ value would have to be 28 h^{-1}. The $K_L.a$ value for any stirred reactor depends upon factors which include the air flow rate, the method of agitation and the stirring rate. The three factors all influence the bubble density and the bubble size which determine the value of a for any system configuration. As the rate of dissolution of oxygen is not rate determining, it is important to ensure that it is above the threshold value below which it would become rate limiting.

5.3.2.2 Bacterial growth

The growth of newly inoculated bacteria in a suitable medium solution (Schlegel, 1986) undergoes changes with time demonstrated by the graph in Fig. 5.1.

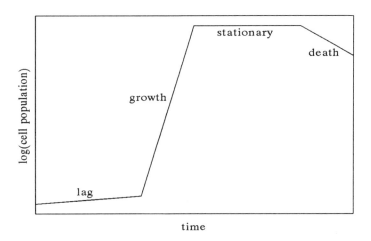

Fig. 5.1 A representation of the four phases of bacterial existence

There are four distinct phases to the growth curve. Initially there is a **lag phase** in which the bacteria are becoming accustomed to their new environment and setting up internal steady states of molecules (enzymes, RNA, ribosomes) essential for cell maintenance and division.

The second phase is the **growth** or **exponential phase** and corresponds to the steeply increasing part of the graph. In the growth phase the bacterial population normally increases exponentially as the bacteria multiply by binary division. An important characteristic of particular bacterial species is its **doubling time**; the time for its population to double under given conditions.

The third phase corresponds to the limitation of the rate of solubilization due to the cessation of growth and is known as the **stationary phase**. In this phase growth ceases because of the depletion of one or more of the essential nutrients or the substrate material. The static bacterial population consists of cells which are still viable. The viable cell population decreases in the fourth phase which is known as the **death phase** in which some cells undergo lysis (disintegration).

The bacteria used in the oxidation of minerals have doubling times between twenty and seventy hours. It is very important that the residence time of the mineral substrate in a continuously operating system should not exceed the appropriate value of the doubling time for the culture used. Failure to operate the system in such a manner leads to the wash-out of the bacterial culture which would cause a reduction of the efficiency of the oxidation initially, followed by the eventual loss of oxidative capacity. In efficient continuously operating systems it is necessary for the conditions to be appropriate for the bacterial culture to be constantly in its growth phase.

5.3.2.3 Attachment of bacteria to mineral surfaces

For the bacterial oxidation of minerals it is essential for the mineral surface to be populated by bacteria. The nature of the attachment process, its reversal, and whether the attachment is temporary or permanent have been, and are continuing to be, subjects of a considerable amount of research.

The attachment/detachment processes are normally dealt with by using the Langmuir approach (Atkins, 1982) of considering the bacteria populating the mineral surface by an adsorption mechanism. If the fraction of the mineral surface already populated by bacteria is represented by ϑ the fraction of available surface is $1 - \vartheta$. The rate of attachment of bacteria to the mineral surface depends upon the bacterial population in the solution, B, and the fraction of available surface. The rate of attachment may be written as:

$$\text{Rate of attachment} = k_a B (1 - \vartheta) \quad (5.11)$$

where k_a is the rate constant for the attachment process. The rate of detachment, for temporarily adsorbed bacteria would depend only upon the fraction of the surface covered by them and could be written as:

$$\text{Rate of detachment} = k_d \vartheta \quad (5.12)$$

where k_d is the rate constant for the detachment process. At equilibrium the two rates are equal and, solving for ϑ, gives the equation:

$$\vartheta = k_a B / (k_a B + k_d) \quad (5.13)$$

In a well established bacterial oxidation reaction proceeding on a continuous basis the value of B is normally high enough to populate the surface of any newly added mineral substrate sufficiently. This is consistent with the value of ϑ tending to unity as the value of B increases. Whether the mineral surface is actually completely covered by bacteria is debatable but is of secondary importance. The significant point is that the establishment of equilibrium, with regard to the populations of bacteria in solution and on the mineral surface, is sufficiently rapid for the process not to be rate determining.

No further description of the knowledge concerning bacterial attachment to surfaces is included for the above reason. It is clear, however, that a low bacterial population could be rate limiting.

5.3.2.4 Aqueous transport of inorganic solutes and their interaction with bacteria

There are three types of inorganic solutes which should be considered; (a) reactants (iron(III) and arsenic(III)), (b) products (iron(II), iron(III), arsenic(III), arsenic(V) and sulfur(VI)) and (c) the nutrient ions which are essential for bacterial growth and maintenance.

Reactant and nutritive solutes are required to diffuse to appropriate sites on the bacterial surface and their consequent adsorption or absorption are important processes with regard to bacterial oxidative activity, growth and maintenance. The desorption and transport of product solutes into the bulk solution should also be considered. A build-up of product species in bacterial sites could produce reaction inhibition. Low concentrations of iron(III) or of any of the nutrient ions could lead to their diffusion rates becoming rate limiting. In normal practice none of these processes is important in the determination of the rate of bacterial oxidation as carried out in stirred reactors. Low concentrations of iron(III) and the nutrient ions could be rate limiting in the static methods of bacterial oxidation (dump, heap, vat and *in situ* leaching).

5.3.2.5 The bacterially catalysed stages

The abiotic chemical oxidations of pyrite and arsenopyrite are considerably slower than their bacterial oxidations (see section 4.7). It appears that true chemical oxidation accounts for only a small percentage of the overall reaction, the figure being in the region of five percent. It is possible, therefore, to identify the reaction stages in which the bacteria participate as catalytic agents. The intimate mechanisms of such stages, i.e. how the bacteria carry out the catalysis, are extremely complicated and not well understood. A discussion of the mechanism of bacterial action is to be found in Chapter 3. For the discussion of a general mechanism the many processes which contribute to the oxidative catalysis of a particular mineral surface are considered as a single stage.

5.3.3 The deduction of a general mechanism for the bacterial oxidation of pyrite/arsenopyrite concentrates

The major evidence for the mechanism of a reaction is its rate law. This expresses the way in which the relevant factors (e.g. concentrations of component reactants and products, surface area of the solid substrates and pulp density, bacterial population on the solid surface and in the bulk solution) affect the rate of the process.

It is helpful, in this and future discussions of mechanism of bacterial oxidation reactions, to distinguish between **primary** processes, which are those occurring at the mineral surface and lead to the oxidation and/or solubilization of the mineral, and **secondary** processes which occur mainly in the bulk of the solution and are concerned with the fates of intermediates.

5.3.3.1 Primary processes

The elucidation of the nature of the primary processes is dependent upon a variety of observations which are dealt with as follows.

(i) The nature of the mineral.

It is accepted that the rate of bacterial oxidation of a mineral depends upon the **nature** of the mineral, since different minerals are oxidized at different rates. Under the same conditions the rate of solubilization of arsenopyrite is around four and a half times higher than that for pyrite (Barrett *et al.* 1988a). The rate of solubilization of a mineral depends also upon the type of semi-conduction exhibited. Samples of sulfide/arsenosulfide minerals may be either n(**n**ormal)-type or p(**p**ositive hole)-type semi-conductors (Barrett, 1991). In an insulator there is a sufficiently large energy gap between the lowest energy vacant band of orbitals and the highest energy filled band to prevent conduction, which depends upon the existence of partially filled bands. If the otherwise vacant lowest energy band is sparsely populated by electrons this gives rise to n-type semi-conduction. If the highest energy band is almost full and so possesses electron vacancies (which may be thought of as positive holes) the semi-conduction is p-type. The n-type minerals are more easily oxidized than the p-type. The n-type have high energy electrons which are more easily removed in the oxidation process whereas in the oxidation of the p-type minerals electrons have to be removed from a band with lower energy. An insulator has completely filled bands, the energy gap between the highest filled band and the lowest energy vacant band being relatively large. The band gap in semi-conductors is smaller than that in insulators and in metals the bands overlap allowing high conductivity to occur. Impurity atoms can impart semi-conduction properties to a material. The substitution of nickel(II), with a d^8 configuration, for iron(II) (d^6) would make the solid electron-rich and increase the n-type semi-conduction. Alternatively, if a sulfur atom (with an s^2p^4 electronic configuration) is substituted by an arsenic atom (with an s^2p^3 electronic configuration) the substance becomes 'electron-deficient' and p-type semi-conduction arises. Differences in the semi-conduction properties of samples of a given mineral can be correlated with differences in their oxidation rates, but conduction is not a major rate determining property. For example, zinc sulfide (an insulator) and nickel sulfide (a metallic conductor) are both bacterially oxidized at rates which are considerably higher than those observed for pyrite and arsenopyrite (both having intermediate semi-conduction properties).

Norman and Snyman (1988) using electron microscopy, observed important differences between the chemical and biologically catalysed oxidations of arsenopyrite and pyrite. The chemical oxidations were carried out with iron(III) (7 g L^{-1} or 125 mM) at a pH value of 1.8 at 20°C. To eliminate any possibility of biological action the solutions contained 100 ppm of thymol to ensure sterility. It was observed that the mineral surfaces had pits and cracks and that after twenty-one to twenty-three days of the treatment the arsenopyrite surface had been totally destroyed. The pyrite surface had been deeply etched but was still recognizable. These observations are consistent with the expected rate differences between the two minerals.

The bacterial leaching was carried out at the same pH value and temperature, the solution containing 9 K nutrient medium (Silverman and Lundgren, 1959). The bacterial culture used contained *T. ferrooxidans*, a sulfur oxidizer (probably *T. thiooxidans*) and *L. ferrooxidans*. The arsenopyrite surface was destroyed after twelve to sixteen days with that of the pyrite showing only slight etching.

Significant differences were observed at the higher magnifications used (x 2200). The bacterial leaching of arsenopyrite progressed from fine cracks, 0.2-0.3 μm wide, to depressions which were 0.8 μm wide and 1-10 μm long (the approximate size of a rod-like bacterium). The pits increased to a sizes of up to 4 μm x 20 μm and eventually merged so that the mineral was totally destroyed. In the same time period the pyrite surface showed only the first stage of this destruction and only in less than twenty percent of the particles.

The chemical leaching produced an evenly distributed superficial etching but the bacterial attack was more localized and led to the eventual destruction of the three dimensional lattice. There was no doubt that the bacteria adhered to the mineral surfaces and that the mode of surface destruction was very different from that produced by the chemical iron(III) attack. The authors do not indicate which substrate was used for the growth of their mixed culture. Whatever it was, it would lead to the inoculum containing a high concentration of iron(III) which could contribute to the bacterial destruction of the minerals.

Using an extreme thermophile culture (*Acidanus brierleyi* DSM 1651), at 70°C, Larsson *et al.* (1991) carried out an ingenious experiment using a reactor with two compartments separated by a membrane which prevented the passage of bacteria but allowed equilibration of the solution phase. One half of the reactor contained a sterile one percent (w/v) suspension of pyrite, the other half contained the culture which derived its energy by oxidizing added iron(II). Any oxidation of the pyrite was due to the iron(III) generated by the bacteria. The experiment showed that the sterile pyrite underwent oxidation at only four percent of the rate at which it oxidized if the culture was allowed to have physical contact (i.e. when bacteria were present in both compartments). Such experiments need to be done with the other cultures used in bacterial oxidation reactions to ascertain whether similar results are achieved.

The conclusions from this section are that the rate of bacterial oxidation depends upon the nature of the mineral and that there is a significant difference, in the mode of attack and in rate, between bacterially catalysed and chemical processes, the latter being relatively very slow. The identification of the products of the primary bacterial oxidation of pyrite and arsenopyrite are central to the understanding of the mechanism of the overall process.

(ii) The role of iron(III) and the identification of intermediates.

The possible participation of iron(III) in the primary process was discussed by Brock and Gustafson (1976) (see also section 3.3.1.5). They showed that the bacteria *T. ferrooxidans*, *T. thiooxidans* and *Sulfolobus acidocaldarius* could reduce iron(III) when growing on elementary sulfur as an energy source. More recently it has been shown (Pronk *et al.* 1992) that under anaerobic conditions *T. ferrooxidans* can derive energy from the oxidation of sulfur by iron(III). This may be the case generally for the bacteria causing the catalysis of the oxidation of refractory minerals and has important consequences. It would be expected that a high iron(III) concentration would be beneficial to the rate of bacterial oxidation and it would mean that it is then less important to establish highly efficient oxygen transfer to the site of primary bacterial oxidation. If this is the case there still has to be sufficient oxygen entering the system to re-oxidize any Fe(II) produced when Fe(III) acts as the oxidant. The total oxygen demand is unaltered by such a consideration.

The chemical reaction of iron(III) with arsenopyrite in acid solution is around six times slower than if suitable bacteria are present (Barrett *et al.* 1990). They report that the chemical reaction is:

$$FeAsS + 11Fe(III) + 7H_2O \rightarrow 12Fe^{2+} + H_3AsO_3 + HSO_4^- + 10H^+ \qquad (5.14)$$

and could be that which best represents the bacterially catalysed primary process. This would imply that iron(II) and arsenic(III) are intermediate products which are oxidized further by secondary processes. These intermediates are the ones described by Shrestha (1988) for the bacterially catalysed reaction with oxygen as the sole oxidant. The chemical dissolution of pyrite is extremely slow.

Additional evidence for the intermediates is furnished by the observations of experiments when the solution bacteria were washed out from an arsenopyrite/pyrite oxidation (Barrett *et al.* 1989a). There was a rapid build-up of iron(II) and arsenic(III) in the solution before the reaction ceased completely. These normally intermediate products were produced by the bacterial oxidation occurring mainly on the remaining mineral surface. The concentration of arsenic(III) eventually reached toxic levels, all reaction ceasing.

Although it is not necessary to add iron(III) to a bacterial oxidation reactor it has been observed (Barrett *et al.* 1989b) that the rates of solubilization of iron and arsenic are roughly proportional to the initially added iron(III) concentration up to a value of around 15.5 g L^{-1} (275 mM) as is shown in Fig. 5.2.

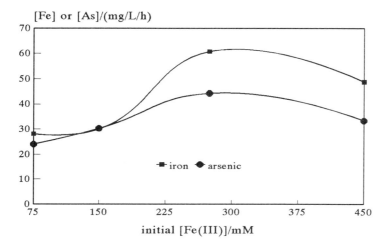

Fig. 5.2 The variations of the rates of solubilization of iron and arsenic with the concentration of added iron(III)

Higher concentrations of iron(III) have little or no effect upon the rate and are possibly inhibitory. This would indicate a saturation effect possibly concerned with adsorption on either the mineral surface or the bacteria (or both). This behaviour is consistent with a Langmuir type of adsorption process in which reactive sites are being occupied by the iron(III) species.

It is possible to conclude from the observations described in this section that at low concentrations of iron(III) (less than 275 mM) the rate of oxidation is dependent upon the concentration of iron(III). It seems likely that it is implicated in the primary process and that iron(II) (released from the mineral surface in addition to that produced as a reduction product of the iron(III)) and arsenic(III) are the intermediate products of the primary process. These observations do not exclude the participation of oxygen as a primary oxidizing agent and it could be that iron(III) and oxygen fulfill different, but essential, roles in the process.

(iii) The dependence of rate upon the bacterial population, particle size and pulp density.

The bacterial population of a given system is dependent upon the total available surface area of the mineral substrate. The available surface area is dependent upon the particle size distribution and the pulp density of the solid substrate. The three factors are interdependent and are discussed together in this section.

The growth of newly inoculated bacteria in a mineral/nutrient slurry shows three phases as demonstrated by the graph of the concentration of solubilized iron against time after inoculation in Fig. 5.3.

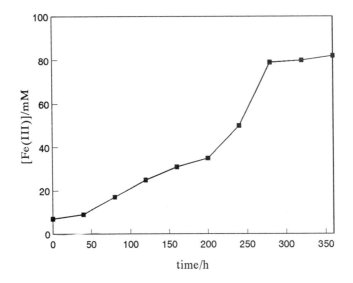

Fig. 5.3 The bacterial solubilization of iron as a function of time

Initially there is a lag phase in which the bacteria are becoming accustomed to their new environment. In this phase growth is minimal although there is some equilibration with the solution components and some slight solubilization of the substrate mineral in the later part. This second part of the lag phase is probably due to the increasing population of the bulk solution by bacteria which is essential to the overall process.

The second phase is the growth phase and corresponds to the almost linear part of the graph. In the growth phase the bacterial population normally increases exponentially and allows sufficient coverage of the mineral surface to maximize the solubilization rate. The linearity of the graph in the growth phase is consistent with the direct bacterial attack on the existing surface, the rate of solubilization falling off only as the mineral substrate is used up and with the onset of the stationary phase. In comparative systems it has been observed (Barrett et al. 1987) that the maximum rate is inversely proportional to the particle size of the substrate for a given pulp density.

The conclusion that can be drawn from the above observations is that there must be an optimum bacterial population present in a bacterial oxidation system to maximize the rate of the primary processes and that the surface area of the mineral is rate determining. For pyrite/arsenopyrite substrates the solubilization rate is optimized by there being a large surface area available and for the iron(III) concentration to be in the region of 275 mM. The optimum particle size is around 45 μm diameter, for the particular mineral concentrate used, and the maximum pulp density consistent with such a particle size is around fifteen percent (w/v) as can be seen from the graph in Fig. 5.4.

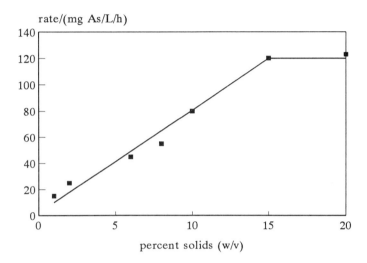

Fig. 5.4 Variation of the rate of solubilization of arsenic with the pulp density of the slurry

Variations in the rates of oxidation of mineral samples from different sources are observed for a given particle size. This indicates that different samples may have different specific surfaces. Both the specific surface and the tortuosity of the mineral sample may affect the specific reaction rate of oxidation.

(iv) Conclusions with regard to the primary process.

With optimum conditions of (a) sufficiently high bacterial population and, (b) an iron(III) concentration of about 275 mM, (c) no more than fifteen percent (w/v) solids density of particles which are -45 μm in diameter and (d) a sufficient supply of oxygen, the primary solubilization of the substrate occurs at maximum rate. The contribution from indirect processes (chemical attack by iron(III)) is negligible and should be discounted in any overall general mechanism. The rate of bacterial oxidation of a mineral is a function of its nature and its surface area, the concentration of iron(III) and the bacterial population. The rate law for the bacterially catalysed oxidation of arsenopyrite/pyrite mixtures may be written in the form:

$$\text{Rate} = f([\text{Fe(III)}], \text{surface area}, \text{bacterial population}) \quad (5.15)$$

Without specifying the form of this functionality (which would be specific to any particular mineral sample) it is possible to conclude that if the conditions are optimum the rate becomes independent of all three factors contained by equation (5.15) and is then constant. The rate is independent of the concentrations of iron(II) and arsenic(III) under circumstances where that of arsenic(III) is sufficiently low to avoid bacterial toxicity.

5.3.3.2 Secondary processes

(i) The primary products.

The primary products of the bacterial oxidation of arsenopyrite are Fe(II), As(III) and S(VI) (although it is possible that some or all of the sulfur goes through the elementary state, the oxidation of the sulfur molecule, S_8, being described as a secondary reaction), all of which are present in the solution phase at the operating pH values of 1-1.5. Iron(II) is the primary product (again elemental sulfur is a possible intermediate) of the bacterial oxidation of pyrite. The iron(II) (Fe^{2+}) and arsenic(III) (H_3AsO_3) are intermediates and as such are oxidized further (to Fe(III) and As(V) respectively) by what may be described as secondary processes.

(ii) The rates of the secondary reactions.

There are no reports of the rates of the secondary reactions as they would occur in bacterial systems. Such data would be obtainable by observing the reductions in the concentrations of iron(II) and arsenic(III) added to a bacterial oxidation reaction operating under steady state conditions. The secondary reactions must all be faster than the primary processes otherwise iron(II) and arsenic(III) concentrations would feature in the rate law. The concentrations of the intermediates are all relatively low (20 mM arsenic(III) and 5 mM iron(II) under normal steady state conditions). This is consistent with their further oxidation rates being more rapid than their production rates.

    A frequently quoted paper (Singer and Stumm, 1970) contains experimental measurements of the rate of the abiotic oxidation of Fe(II) to Fe(III) by dissolved oxygen. They report that the oxidation of pyrite in the presence of iron(III) is independent of the presence of oxygen and, at a pH value of unity, fifty percent of the iron(III) initially present was reduced to iron(II) by pyrite in fifty minutes. In the absence of iron(III) under similar conditions there was no observable reaction of the pyrite after one week. Separate experiments showed that the rate of oxidation of iron(II) by oxygen was independent of pH (at pH values lower than three) and was first order in the concentration of iron(II) and oxygen partial pressure. The rate constant was reported to be 1×10^{-7} atm^{-1} min^{-1} which, under normal atmospheric conditions, indicates a half-reaction time of about 24000 days for the oxidation. If the abiotic oxidation of pyrite were to be the only one operating in mines then acid mine drainage would be no problem.

It has been reported (Silverman and Ehrlich, 1964) that microorganisms in acidic mine water catalyse the reaction by a factor larger than 10^6. This would indicate a half-time for the oxidation of iron(II) of about thirty minutes which would indicate that the iron(III) oxidation of pyrite would become rate determining (the slowest step) in the presence of catalytic bacteria. In spite of this Singer and Stumm reach the doubtful conclusion that the iron(II) oxidation reaction is still rate determining. If this were the case the overall rate of solubilization would depend upon the iron(II) concentration and would not depend upon the iron(III) concentration and the pulp density. There is no doubt, however, that the bacterially catalysed oxidation of iron(II) to iron(III) is a major secondary process leading to the replenishment of the iron(III) in the bacterial oxidation system. The bacteria *T. ferrooxidans, L. ferrooxidans, S. thermosulfidooxidans,* and components of the moderate thermophilic genera (*Sulfolobus* and *Acidanus*) have the capacity to catalyse the oxidation of aqueous iron(II), the oxidizing agent necessarily being oxygen.

There is no evidence that dissolved oxygen, even in the presence of bacteria, can oxidize arsenic(III) to arsenic(V) at measurable rates. The thermodynamic feasibility of the homogeneous chemical oxidation of arsenic(III) to arsenic(V) by iron(III) is discussed in section 4.6.5. It has been reported (Barrett *et al.* 1989b) that the reaction does occur in the presence of a culture growing on pyrite. It appears that the reaction is catalysed by both pyrite and chalcopyrite surfaces and it is possible that the bacteria have a role (Barrett *et al.* 1991) in conditioning the surface. There is evidence (Silver *et al.* 1989) that bacteria do not have the capacity to derive energy from the reaction but their presence is certainly necessary for the reaction to proceed efficiently. The surface of arsenopyrite does not seem to have an appreciable capacity to catalyse the oxidation of arsenic(III). This observation is consistent with the suggestion that if iron(III) participates in the primary oxidation process, such reaction will be kinetically preferred (i.e. it is the faster reaction) to the Fe(III)/As(III) interaction. The kinetic preference is reversed in the case of the pyrite surface which is more resistant to bacterial oxidation than is arsenopyrite.

An alternative possible explanation for the catalytic effect of pyrite is that, in its bacterial oxidation, some of the initially released iron(II) does not enter the bulk solution, but becomes oxidized to iron(III) by the bacteria on the pyrite surface. This is equivalent to the primary release and secondary oxidation of iron(II) occurring concurrently. If this is the case the iron(III) produced is likely to be initially formed as a sulfato-complex as indicated by the speciation diagram in Fig. 4.4. If an arsenic(III) species entered this system it would be oxidized to arsenic(V) (the potentials concerned with this reaction are discussed in section 4.6.4). The arsenic(V) would reach its equilibrium state by acting as a ligand to an iron(III) species in the bulk solution.

(iii) Population of bacteria in the solution phase.

From the considerations of the previous section it is possible to conclude that it is essential that there should be should be a high enough population of bacteria in the bulk solution to enable the secondary oxidation reactions (of iron(II) to iron(III) and arsenic(III) to arsenic(V)) to occur at reasonable rates. Direct evidence for this (Budden and Spencer, 1990) arises from measurements of the variation of bacterial population with time in a bacterial oxidation reaction. Their data, illustrated in Fig. 5.5, show the increase in the number of bacteria in solution as the reaction proceeds.

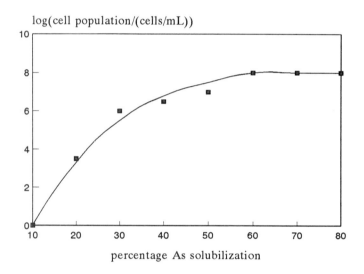

Fig. 5.5 Variation in the bacterial population in solution with the percentage of arsenic solubilized

(iv) The secondary oxidation of elemental sulfur.

Any elemental sulfur produced in the primary stage of bacterial oxidation is oxidized further to the (VI) state by components of the culture which can derive energy from such a reaction. The subject has been reviewed comprehensively by Kelly (1988).

5.3.4 Mechanistic conclusions

The overall reactions representing the bacterially catalysed oxidation of arsenopyrite/pyrite mixtures may be classified in terms of primary and secondary processes. The main features of the mechanism of the bacterial oxidation of arsenopyrite/pyrite mixtures are outlined in the diagrams shown in Figs. 5.6, 5.7, 5.8 and 5.9. The transport of oxygen and hydrated protons into the cell is omitted from the diagrams. Figs. 3.3 and 3.6 are the diagrams representing the primary oxidation of pyrite. The curved arrows represent chemical changes, the straight ones indicating the direction of electron transport.

The primary processes include the bacterially mediated oxidation of the arsenopyrite/pyrite substrate by oxygen and/or iron(III) to give arsenic(III) and sulfur(VI) as products, with the solubilization of the iron(II) content. Any iron(III) taking part as an oxidant is reduced to iron(II). The primary processes may be described by the equations:

$$\text{FeAsS} \xrightarrow[\text{bacteria}]{\text{oxygen/iron(III)}} \text{Fe(II)} + \text{As(III)} + \text{S(VI)} \tag{5.16}$$

$$\text{FeS}_2 \xrightarrow[\text{bacteria}]{\text{oxygen/iron(III)}} \text{Fe(II)} + 2\text{S(VI)} \tag{5.17}$$

Both of these overall primary reactions have complex stoichiometries, are essentially complicated, and must occur in a number of stages. These will be fully elucidated ultimately but, with regard to plant operation, it is only necessary to know that their rates depend upon there being a sufficiently high bacterial population, a reasonably high concentration of iron(III) and suitable aeration. The diagram shown in Fig. 5.6 refers to the primary oxidation of arsenopyrite with oxygen as the sole oxidant. Fig. 3.3 is the equivalent diagram for pyrite oxidation.

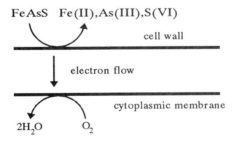

Fig. 5.6 A diagrammatic representation of the primary bacterial oxidation of arsenopyrite with oxygen as the sole oxidant

The diagram shown in Fig. 5.7 includes the possible participation of iron(III) as an oxidant of arsenopyrite in addition to oxygen. Fig. 3.6 is the equivalent diagram for pyrite oxidation.

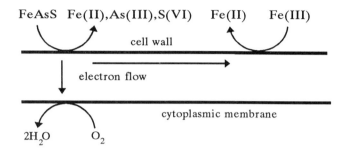

Fig. 5.7 A diagrammatic representation of the primary bacterial oxidation of arsenopyrite with oxygen and iron(III) acting as oxidants

There is a possibility that the primary oxidation could occur anaerobically, such a process being represented by the diagrams shown in Figs. 3.6 and 5.7, but with the omission of the half-reactions in which oxygen is reduced to water. Such processes would take place at the cell wall and would not require electron transport across the cytoplasmic membrane. They would be relevant to systems in which the air supply had failed or was insufficient for bacterial cell requirements.

The iron(II) and arsenic(III) produced in the primary reactions are oxidized to iron(III) and arsenic(V) in secondary bacterially catalysed processes. Oxygen must be the oxidant for iron(II). Iron(III) is the oxidant for arsenic(III), with the pyrite surface also participating. The catalytic function of the solid surface may be to act as an adsorbent for the reactants in the classical Langmuir manner. Another possibility is that iron(III), produced by the bacterial oxidation of pyrite, in the form of a sulfato-complex (because of the local absence of arsenic) is the active oxidant. If this is the case the effect of the pyrite may be explained. The iron(III) sulfato-complexes are potentially more powerful oxidizing agents than any of the arsenato-complexes. Additional evidence for such a conclusion is furnished by the observation that chalcopyrite also catalyses the oxidation of arsenic(III) in bacterial oxidation reactions.

The secondary processes are best represented by the following equations.

$$Fe(II) \xrightarrow[\text{bacteria}]{\text{oxygen}} Fe(III) \qquad (5.18)$$

$$As(III) \xrightarrow[\text{bacteria}]{\text{iron(III)/pyrite}} As(V) \qquad (5.19)$$

The diagram in Fig. 5.8 refers to the secondary oxidation of iron(II) which is catalysed by bacterial cells which occur mainly in the bulk solution of the reacting system. The oxidation of the iron(II) takes place in the periplasmic space as indicated by the position of the curved arrow.

Sec. 5.3]	The mechanism of bacterial oxidation reactions	121

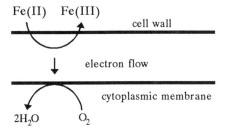

Fig. 5.8 A diagrammatic representation of the secondary oxidation of iron(II)

The diagram in Fig. 5.9 represents a possible mechanism of the secondary oxidation of arsenic(III). The primary oxidation of the pyrite causes the release of iron(II) as indicated by the diagram of Fig. 3.3. The iron(II) may be released into the bulk solution or it may be oxidized immediately without leaving its initial environment. If the latter event occurs the sulfatoiron(III) ion produced could act as an oxidant for any arsenic(III) in the solution. This would imply, as is indicated in Fig. 5.9, that the oxidation of the arsenic(III) occurs in that part of the solution which is in the vicinity of the cell wall of a bacterium which is catalysing pyrite oxidation.

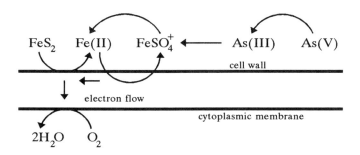

Fig. 5.9 A diagrammatic representation of the secondary oxidation of arsenic(III)

There is a possibility that the oxidation of arsenic(III) could take place on the cell wall, but there is little evidence for this currently. The diagram contains the implication that the cell responsible for the arsenic(III) oxidation also participates in the oxidation of pyrite (as opposed to arsenopyrite). This provisional mechanism is based upon the experimental observations described above and includes only the iron and arsenic pathways. This view of the mechanism will be modified and refined as further experimental results become available.

5.3.4.1 Comments on the previous mechanistic conclusions

The mechanism deduced from all the currently available evidence is more detailed than the previous descriptions which are summarized in section 5.2. Some comments upon the latter are included in this section.

The reaction scheme for the bacterial oxidation of pyrite has been represented by equations (5.2-5.5) and correctly identifies iron(II) and sulfur(VI) as the products of the 'direct' bacterial process (equation 5.2). Equation (5.3) is equivalent to the proposed secondary process represented by equation (5.18). Equation (5.3) which is allegedly the 'indirect' oxidation of pyrite is not correct. The abiotic chemical oxidation of pyrite is a very slow process in comparison to the bacterially catalysed reaction and must play a very minor role in the total process. It would appear that the primary bacterial oxidation of pyrite is best represented by equation (5.17) in which a combination of oxygen and iron(III) attacks the mineral surface. It is possible that the elemental state (sulfur(0)) contributes to the sulfur oxidation pathway as an intermediate. Any sulfur(0) produced as an intermediate is possibly a participant in the detailed mechanism of reactions (5.2) or (5.17) rather than being produced by the slow chemical process. Equation (5.5), representing the bacterially catalysed oxidation of sulfur(0) could be a part of the primary process although the independent oxidation of the element is well established.

Equation (5.7) is a correct expression of the products of the primary bacterial oxidation of arsenopyrite (again the sulfur oxidation pathway may include the elemental state as an intermediate). Equation (5.8) expresses the chemical changes correctly but there seems little doubt that the reaction belongs to the secondary category and is bacterially and surface catalysed.

The terms 'direct' and 'indirect' are inadequate for describing pathways in the bacterial oxidation of minerals and their use should now be discontinued. This was suggested in 1979 (Kelly *et al.*) but the terms have nevertheless persisted in the literature. Bacterially catalysed oxidations of minerals may be discussed more accurately in terms of the 'primary' and 'secondary' processes as defined in the above sections.

5.4 CONSEQUENTIAL PREDICTIONS

Confidence in any scientific theory is established by its ability to predict the behaviour of hitherto untried systems. If a prediction does coincide with subsequent observation it must not be construed as 'proof' of the theory. Theories develop as a result of observations which are contrary to those what would be predicted (Popper, 1959). The theory should either be consistent with new experimental data or undergo modification to be so. The currently proposed mechanism for the bacterially catalysed oxidation of arsenopyrite/pyrite mixtures is at an early stage and will no doubt be modified and refined as further data is generated and published. The purpose of including this section is to help to formulate the observable consequences of plant breakdown and misfunction and to list any danger signals. It also signifies the relevance of mechanistic studies. The observable parameters which are normally used to monitor and control plant performance are the pulp density of the slurry, the values of pH and E (electrode potential) of the slurry solutions and the concentrations of iron(III) and arsenic(V), some or all of which may be estimated on a regular basis. It is also possible to monitor the concentrations of iron(II) and arsenic(III), and the bacterial population (in the solution phase and on the solid surface). The details of the essential analytical methods are the subject of Chapter 9.

The parameters which may be used for monitoring and controlling the process of bacterial oxidation are discussed below. The more important of these parameters, at the plant level, are discussed further in Chapter 6.

5.4.1 pH

In general the addition of fresh solid substrate tends to cause the pH of a slurry to increase. This may be due to some basic content (carbonates, oxides) or to the ion-exchange characteristics of the silicates in the gangue fraction. The effect of substrate on pH should be studied on test material so that some upper limit upon the feed rate is established. This would be in order to avoid the pH of the slurry attaining a value at which the undesirable precipitation of iron(III) compounds may occur. With basic substrates there is a possibility that the local pH (rather than the bulk average value) would be above such a limit temporarily when the solid was first added to the slurry. It is possible that jarosite formation could occur by such a method and affect the rate of bacterial oxidation by partial protection of the mineral surface. Other materials may be precipitated (as are discussed in Chapter 7) which would only slowly redissolve when the optimum pH value of the system was re-established.

As bacterial oxidation proceeds in a batch reactor there is a general decrease of pH due to the production of sulfur(VI) (sulfuric acid). If the process is carried out in a continuous manner, as is normal on plant scale, the steady state pH value depends upon the opposing effects of the sulfuric acid production and the addition of basic substrate. The observed steady state value of the pH of the slurry under normal conditions could be between 0.5-1.5. The practice of increasing the pH by the addition of lime or limestone is inadvisable. It could lead to local pH values above two to three thus producing precipitates which could interfere with the reactive mineral surface. The presence of calcium ions could lead to the removal of the phosphate ion which is one of the essential ingredients for bacterial growth.

5.4.2 Iron(III) concentration

The iron(III) concentration in a slurry under steady state conditions should be around 0.3 M. Iron(III) participates in the primary process and in the secondary oxidation of arsenic(III). Lower values of the concentration of iron(III) lead to lower overall oxidation rates. Values greater than 0.3 M should be avoided as there is evidence that the primary oxidation process is inhibited by such concentrations. In terms of enzyme kinetics this would be regarded as product inhibition but it could be that there is an ideal balance between the bacteria and the iron(III) species such that an excess of the latter is either deleterious to the former's functionality or to its population on the mineral surface. Iron(III) is certainly required for the necessary oxidation of the arsenic(III) produced in the primary stage.

5.4.3 Arsenic(V) concentration

Arsenic(V) is a final product of the secondary oxidation of the initially produced arsenic(III) from the primary oxidation of arsenopyrite. There is no evidence to suggest that it takes any further part in the mechanism of the process. Its normal steady state concentration is around 0.2 M (15 g L^{-1}) although levels of 0.27 M (20 g L^{-1}) have been observed in successfully operating plants. The bacterial cultures used generally adapt to such high arsenic(V) levels and to an adapted culture they are not toxic. Arsenic(V) toxicity is only a danger in the initial stages of adaptation and is a laboratory problem since cultures used in plant operation have already been adapted to resist appropriate levels. If it became necessary to change the mineral substrate to one with a higher arsenic content than was previously used there might be problems with toxicity and such a change should be carried out after appropriate laboratory testing.

5.4.4 Arsenic(III) concentration

Arsenic(III) is a primary product of the bacterial oxidation of arsenopyrite and is normally oxidized to arsenic(V) in a secondary process involving the bacteria, iron(III) and a pyrite surface. The normal level of arsenic(III) in a reactor is around 20 mM (1.5 g L^{-1}) which indicates that its rate of oxidation is around ten times faster than its rate of production. In a well adapted culture such a concentration poses no toxicity to the bacteria. Arsenic(III) is approximately three times more toxic than arsenic(V) and a sudden increase in the arsenic(III) concentration could be very damaging to a culture. The proposed mechanism would predict an increase in the concentration of arsenic(III) if the secondary oxidation process, taking place in the bulk solution, were to slow down or stop. This possibility would occur if conditions allowed the wash-out of the solution containing the bacteria responsible for the secondary processes. Such a condition could result from inefficient agitation of the slurry. This would cause settlement of the solids possibly resulting in a relatively inefficient transfer of solids to the next stage of the plant compared to the transfer of solution. The primary oxidation would continue the production of arsenic(III) the concentration of which would build up to a toxic level and which would damage or disable the culture.

5.4.5 Iron(II) concentration

Iron(II) is a primary and secondary product of the oxidation process. It is released from the mineral in the primary process and is a primary reduction product of any iron(III) which takes part in that stage of the process. It is also a product of the secondary oxidation of arsenic(III) by iron(III). The normal steady state concentration of iron(II) is around 5 mM (0.28 g L^{-1}), the level indicating that the rate of its oxidation is about sixty times higher than the rate of its formation (by all processes). It poses no threat to the bacteria and no inhibition has been observed. The concentration of iron(II) would build up under the same circumstances as those described for that of arsenic(III). Its concentration is more easily monitored than that of arsenic(III) and could serve as an easier indication of solution bacterial wash-out.

5.4.6 Electrode potential

The electrode potential, E (sometimes referred to by the symbol, E_h), developed between a standard mercury-calomel or silver-silver chloride electrode and a platinum electrode which are both immersed in the reactor solution is often used to indicate the oxidizing power of the system. The value of E with respect to the hydrogen electrode is obtained from the experimentally observed value by subtracting the value of the potential of the standard electrode (e.g. 0.244 V for a saturated potassium chloride/mercury-calomel electrode at 298 K). In some publications it is not clear whether this operation has, or has not, been carried out. The experimental electrode pairs used in practice react only to the relative concentrations of two oxidation states of an element when the electrode reactions (electron gain or loss) take place reversibly. In practice this means iron(III) and iron(II). The other pair of oxidation states, (As(III) and As(II)), do not interact reversibly with electrode surfaces since their redox reactions are very slow. This is generally true for oxoanions or any molecules which differ in the number of oxygen atoms bonded to the central ion. The redox process in such cases inevitably requires central atom-oxygen bond-breaking as well as electron transfer. These so-called group transfer reactions are very slow compared to the more straightforward electron transfer reactions which are typified by the Fe(III)/Fe(II) redox interaction.

The observed value of E is thus an indication of the value of the [Fe(III)]/[Fe(II)] ratio. It offers no indication of the value of the [As(V)]/[As(III)] ratio as is seemingly sometimes believed. As such it is a reasonably good indication of the iron(II) level, a low value of E being associated with a high iron(II) concentration. Bacterial oxidation systems which are operating optimally have low iron(II) levels which cause the observed value of E to be high. Low values of E are reportedly associated with poor bacterial oxidation performance. It is sometimes implied that there is a causal mechanistic connexion between the low value of E and the poor system performance. This is rather unlikely and the reverse is more probable. If, for some reason, the secondary processes are deleteriously affected so that iron(II) builds up in the solution this would cause the value of E to decrease.

5.4.7 Pulp density (solids density)

The normal value of the pulp density in a continuously operating bacterial oxidation reactor stage is twenty percent (w/v) solids, the rate of oxidation being proportional to the pulp density up to that value. Any increase in solids above that figure results in no further change in the rate of the reaction. The optimum value must reflect the balance between the maximum reactive mineral surface and the bacteria (on the surface and in the bulk solution) which can be maintained at their maximum growth by that surface. If the pulp density exceeds twenty percent too high a fraction of the bacteria is associated with the surface and too low a fraction remains to optimize the secondary processes. A sudden increase in the pulp density to an extent which reduces the solution population to a low level has the effect of stopping the secondary processes and the arsenic(III) level builds up rapidly causing bacterial toxicity which results, at best, in reducing the culture to its lag phase. The pulp density should not be allowed to rise above the twenty percent level and the rate of addition of fresh mineral substrate should be consistent with this. It is not advisable to add the fresh substrate infrequently in large amounts but to add it either continuously or frequently in small amounts.

5.4.8 Bacterial population

The chemical monitoring of bacterial population upon the mineral surface and in the bulk solution is a relatively difficult and lengthy procedure. It would be used to show the the populations in the two phases were satisfactory or otherwise. In plant operations it is practical to use microscopy to occasionally check the bacterial population level. Only in investigating problems would it be necessary to use the chemical procedures.

5.5 REFERENCES

Atkins, P.W. (1982) *Physical Chemistry.* 2nd edn, Oxford University Press, p.1023.
Barrett, J. (1991) *Understanding Inorganic Chemistry.* Ellis Horwood. p.122.
Barrett, J. (1992) *unpublished observation.*
Barrett, J., Hughes, M.N., Nobar, A.M., O'Reardon, D.J. & Poole, R.K. (1987) *unpublished work.*
Barrett, J., Ewart, D.K., Hughes, M.N., Nobar, A.M., O'Reardon, D.J. & Poole, R.K. (1988a) *R & D for the Minerals Industry, Kalgoorlie, 1988.* Western Australian School of Mines, p.275.
Barrett, J., Hughes, M.N., Nobar, A.M., O'Reardon, D.J. & Poole, R.K. (1988b) *unpublished work.*
Barrett, J., Ewart, D.K., Hughes, M.N., Nobar, A.M. & Poole, R.K. (1989a) *unpublished work.*

Barrett, J., Ewart, D.K., Hughes, M.N., Nobar, A.M. & Poole, R.K. (1989b) *Biohydrometallurgy - 89, Jackson Hole, 1989*. Salley, J., McReady, R.G.L & Wichlacz, P.L. (eds), p.49.

Barrett, J., Hughes, M.N. & Russell, A. (1990) *Randol Gold Forum, Squaw Valley, 1990*, p.135.

Barrett, J., Ewart, D.K., Hughes, M.N. & Poole, R.K. (1991) *Biohydrometallurgy - 91, Portugal, 1991, in press*.

Brock, T.D. & Gustafson, J. (1976) *Appl. Environ. Microbiol.*, **32**, 567.

Budden, J.R. & Spencer, P.A. (1990) *Proceedings of the TMS Annual Meeting, Anaheim, 1990*, D.R.Gaskell (ed.) p.295.

Kelly, D.P. (1988) *The nitrogen and sulfur cycles, S.G.M. Symposium, 1988*, Cole, J.A. & Ferguson, S. (eds) **42**, 65.

Kelly, D.P., Norris, P.R. & Brierley, C.L. (1979) *Microbial Technology: Current State, Future Prospects, 29th Symposium of the Society for General Microbiology, Cambridge, 1979*, Bull, A.T., Ellwood, D.C. & Ratledge, C. (eds) Cambridge, p.270.

Larsson, L., Olsson, G., Holst, O. & Karlsson, H.T. (1991) *Biohydrometallurgy-91, Portugal, 1991, poster presentation*.

Norman, P.F. & Snyman, C.P. (1988) *Geomicrobiology J.*, **6**, 1.

Panin, V.V., Karavaiko, G.I. & Pol'kin, S.I. (1985) *Biogeotechnology of Metals. Moscow, 1985*, Karavaiko, G.I. & Groudev, S.N. (eds), p.197.

Popper, K.R. (1959) *The Logic of Scientific Discovery*. 14th impression, Unwin Hyman, 1990, p.33.

Pronk, J.T., Liem, K., Bos, P. & Kuenen, J.G. (1992) *Appl. Environ. Microbiol., in press*.

Rossi, G. (1990) *Biohydrometallurgy*. McGraw-Hill, p.285-320.

Russell, B. (1946) *The History of Western Philosophy*. George Allen and Unwin Ltd., p.494.

Schlegel, H.G. (1986) *General Microbiology*. 6th ed., Cambridge University Press.

Shrestha, G.N. (1988) *Australian Mining* p.48.

Silver, S., Nucifora, G., Chu, L. & Misra, T.K. (1989) *Trends Biochem. Sci.*, **14**, 76.

Silverman, M.P. & Ehrlich, H.L. (1964) *Advan. Appl. Microbiol.*, **6**, 153.

Silverman, M.P. & Lundgren, D.G. (1959) *J. Bacteriol.*, **77**, 642.

Singer, P.C. & Stumm, W. (1970) *Science* **167**, 1121.

6

Application of bacterial oxidation technology

6.1 INTRODUCTION

This chapter deals with the various methods of applying bacterial oxidation technology as full scale industrial processes. It includes a discussion of some fundamental operating conditions which are independent of the chosen method and this is followed by a detailed discussion of the various possible methods of application of the technology and the dependence of these methods upon the particular processes being carried out (class I metal liberations or class II/III solubilizations).

There are five general methods of carrying out bacterial oxidation reactions as full scale operations. These involve the use of (i) agitated reactors, (ii) prepared heaps, (iii) dumps, which may or may not have been constructed with extraction in mind, (iv) vats (lined ponds in which the crushed mineral is flooded with water) and (v) *in situ* or in-place leaching (where the ore body undergoes bacterial oxidation and produces a leachate containing the metal values). An outline of the methods is included in Chapter 2, a general flow-sheet being shown in Fig. 2.1.

6.2 FUNDAMENTAL OPERATING CONDITIONS

There are several fundamental operating conditions that may be used to control a bacterial oxidation process, regardless of the method of application or the sulfide mineral being treated. These operating conditions vary in their relative importance with the application being considered and in their ease of control (depending upon the method of bacterial oxidation adopted), but all must be maintained within certain limits to achieve optimum plant performance.

The most important operating conditions are:

(i) quantity of sulfide mineral present (pulp density and particle size distribution in the case of agitated tank methods for mineral concentrates, percentage mineral content and particle size distribution in the other cases for ore treatment),

(ii) aeration,

(iii) acidity,

(iv) temperature,

(v) nutrient supply, and

(vi) culture growth.

In some cases other factors, e.g. the supplementation of the natural carbon dioxide in air, may be used to control the culture growth. Special measures to promote culture growth can be necessary in applications where one or more of the major parameters cannot be adequately controlled.

A consideration of some importance is the recognition that bacterial cultures employed in bacterial oxidation processes in non-sterile environments are likely to contain more than one species, often because of the presence of indigenous organisms in the mineral matrix to be treated. This is particularly true of any method which processes ore. Freshly produced concentrates, because of their methods of production by the use of potentially toxic gathering and frothing reagents, are unlikely to be contaminated by indigenous bacteria. If they are washed (to remove the flotation reagents) and stored before being bacterially oxidized they are likely to attract iron/sulfur oxidizing bacteria from their surroundings in the course of time. Alteration of any of the factors that might be used to control or improve the process can affect the balance of the different organisms present (because they have different growth rates), sometimes with an unexpected deleterious effect on the process performance.

6.2.1 Quantity of sulfide material

In a tank reactor system the pulp density of the concentrate is the primary rate determining factor as is discussed in section 5.3.3.1 (iii). Under normal continuously operating conditions the bacterial population is dependent upon the pulp density of the substrate used, with its inherent particle size distribution. The optimum pulp density for any particular substrate is estimated during the necessary laboratory scale amenability testing which precedes any plant operation. The optimum operating value of the pulp density is then attained by fine tuning at either the pilot and/or full scale plant stages. In practice the value indicated by laboratory amenability tests is a good indication of that required for larger scale operations.

In any given case, using the same substrate mineral concentrate, the optimum pulp density depends primarily upon the particle size distribution. Any changes in the distribution affect the pulp density to be used, lower particle sizes demanding lower values of the operating pulp density.

An additional consideration in deciding the optimum sulfide mineral concentration is the final operating concentration of reaction products in the solution phase of the slurry and the relative toxicity of these dissolved species to the bacterial culture being employed in the process.

After the initial inoculation of the appropriate culture, the bacterial population in a reaction system normally grows to a level which is sustainable by the conditions and the quantity of sulfide mineral present. The population of bacteria in the solution phase should be normally maintained at a level which is sufficient to ensure the coverage of any new sulfide surfaces that become available without reducing the solubilization rate of the sulfide minerals present and without severely depleting the number of cells in suspension. Freshly added mineral concentrate adsorbs bacteria from the bulk solution phase so that the primary oxidation can take place. It is important that the rate of addition of new mineral should not deplete the population of bacteria in the solution phase. If the bacterial numbers do not remain in appropriate balance between the solution and solid phases there is a possibility that all the bacteria in the system may be attached to the surface of the sulfide mineral and may flow into the next stage of the plant, leaving insufficient bacteria to sustain the reaction in the initial stages. This phenomenon (washout of bacteria) can be mistakenly ascribed to the disablement or death of the bacterial culture. The greater the surface area of the sulfide mineral, which is controlled by particle size and by the sulfide content of the slurry, the higher is the risk of depletion of the numbers of bacteria in the initial stage of the bacterial oxidation process. As is discussed in section 5.4.7, a slurry containing only surface-adsorbed bacteria would revert to a lag phase bacterial growth condition and would contribute to a lower solubilization rate in the system. Proper selection of grinding range size and pulp density in the slurry can prevent the excessive loss of bacteria from the initial stages of bacterial oxidation.

Most bacterial cultures have a limiting concentration of metal ions that can be tolerated while maintaining maximum bacterial oxidation performance. This limit varies according to the culture used in the process but it has been reported (Brierley and Brierley, 1986) that moderately thermophilic cultures are generally more resistant to high metal concentrations than the more commonly used *Thiobacillus*-based mesophilic cultures. With continued exposure to increasing concentrations of a toxic component of the system the culture being employed can usually, by a process of adaptation, become sufficiently resistant so that for practical purposes the toxicity of that component is no longer a problem.

Some of the dissolved elements considered to be particularly toxic to bacterial cultures are copper, nickel, mercury, silver, molybdenum and arsenic. The toxic effect of such dissolved elements can be controlled by proper selection of solids density in the mineral slurry so that when the level of oxidation of the sulfide mineral reaches the optimum level the concentration of the dissolved species remains below that which would affect the performance of the bacterial culture. As processing proceeds the adaptation of the culture to its environment allows the use of higher pulp densities.

The amount of metal that can be dissolved in the solution phase of the system being treated is also partly determined by the solubility of ions of the metal in the presence of the other dissolved species (reactants, products and nutrients) at the operating temperature of the process. For example, if the solubility of a metal salt is exceeded in the bacterial oxidation process a precipitate will form, perhaps removing the valuable metal in an insoluble form that cannot easily be recovered at a later stage, or forming a coating on the sulfide mineral, or blocking the solution flow which could prevent further reaction. As a second example, in a process which solubilizes arsenic, it is important to prevent uncontrolled precipitation of arsenic compounds before the solution containing the dissolved metals is separated from the reacted solids so that an environmentally stable iron-arsenic precipitate can be produced. Precipitation within the slurry cannot be easily controlled and reduces the probability of producing a stable precipitate for safe disposal.

The sulfidic mineral content of the other methods of bacterial oxidation (heaps, dumps and vats) cannot be controlled as optimally as in the case of agitated tank operations. The effects of particle size distribution upon the other crucial factors are considered separately in sections 6.3.2 (heaps), 6.3.3 (dumps) and 6.3.4 (vats). In the static methods the concentrations of reactants, products and nutrients (if any are added) are normally too low to cause any toxicity problems.

6.2.2 Aeration

Aeration of the solution (used in heap, dump or vat leaching methods) or slurry (used in agitated tank methods) is important in all bacterial oxidation processes using obligately aerobic acidophilic bacterial cultures. If the oxygen concentration falls to low levels, less than 0.5 to 1.0 mg L^{-1} for processes carried out in stirred vessels, the culture will normally revert into its lag phase and the bacterial process will stop. A lack of carbon dioxide restricts the culture growth and could limit the rate and amount of reaction of the sulfide mineral.

There have been a large number of conflicting reports on the level of aeration required for bacterial oxidation processes. In some cases stoichiometric equations have been written for the chemical reactions and the amount of aeration calculated using oxygen transfer information from waste water treatment. In other reports the level of aeration has been based upon actual measurements, often in situations where significant chemical oxidation is likely to occur. At the present time there is no definitive publication where the oxygen and carbon dioxide requirements have been measured under conditions where only bacterial oxidation processes are occurring.

In heap, dump and vat leaching operations no enrichment of the natural levels of oxygen and carbon dioxide in the water employed is deemed necessary apart from that achieved by spraying water over their surfaces (in the cases of heaps and dumps).

Control of the content of the solution running through the mine is not possible in the case of *in situ* leaching. Although the supply of the essential gases may very well be rate limiting, enrichment would be too expensive to use in these methods.

Experience to date implies that oxygen and carbon dioxide transfer from the gaseous phase is not a problem with agitated reactor systems although it may restrict the rate of the bacterial oxidation reactions in the static methods.

6.2.3 Acidity

The levels of acidity (measured in terms of pH values) reported for growth of the cultures used in bacterial oxidation processes cover a much greater range than can be employed in practical applications. The pH range employed in practical bacterial oxidation processes is determined largely by the bacterial culture used, the sulfide sample being treated and the nature of the bacterial oxidation processing method adopted. Bacterial cultures have a limited range of acidity over which bacterial oxidation will proceed. Most sulfide ores and concentrates contain iron which can precipitate at higher pH values and retard or prevent the bacterial oxidation process, and processes such as heap and dump leaching make precise control of acidity difficult.

Optimum pH ranges for growth of *Thiobacillus ferrooxidans* and *Thiobacillus thiooxidans* are reported (Kuenen and Tuovinen, 1981) to be 3.0-5.5 and 2.0-4.5 respectively. In practice, conventional *Thiobacillus*-based cultures have been applied at pH values in the range 1.5-3.0, most commonly at about pH 1.8-2.0. This range is not necessarily used because it is the optimum range for bacterial growth. The upper pH limit is set by the level of acidity at which iron salts dissolved in the bacterial oxidation process are less liable to precipitate as hydrated iron oxides (or, if arsenic is present, as iron(III)/arsenic(V) compounds). The lower limit is set by the highest level of acidity which the bacterial culture will tolerate and still perform the bacterial oxidation.

The static methods exploit the low pH of acid mine water (if that is the source of the water employed) or that of the water used, the pH being lowered as the solution percolates through the heap or dump. In the case of vat leaching it would be possible to lower the initial pH by addition of a suitable amount of sulfuric acid.

6.2.4 Temperature

The operating temperature of a bacterial oxidation process cannot always be controlled. The optimum temperature range is normally the one which results in the fastest oxidation of the sulfide mineral. The temperature is not always the optimum value for the growth of the bacterial culture. Each bacterial culture has a preferred temperature for growth and oxidation, but other factors such as chemical leaching, the type of bacterial process, the local climate and ease of temperature control, influence the operating temperature for any particular bacterial process.

Thiobacillus-based cultures frequently operate in agitated tanks at temperatures between 35-40°C to take advantage of the chemical leaching at the higher temperatures and to reduce the cooling requirements when heat might be generated in the process. The same cultures can act successfully in heap and dump leaching operations at temperatures which may be 10-15°C lower. Vat leaching temperatures are dependent upon the variable ambient temperatures at the location.

6.2.5 Nutrients

The standard nutrient solution reported in the literature for isolation and growth of *Thiobacillus ferrooxidans* cultures is the so-called 9K medium (Silverman and Lundgren, 1959) or a suitable derivative of that solution. The solutions contain nitrogen as an ammonium salt, phosphorus as a potassium salt of phosphoric acid, magnesium as magnesium sulfate and iron as iron(II) sulfate. Other salts that are sometimes included are calcium nitrate or calcium chloride. The pH of the nutrient solution is normally adjusted to a final value of between 1.3-3.5.

Not all of these salts are required in practical leaching situations because some essential elements are present in either the water used or the mineral matrix being treated. Iron is frequently omitted in sulfide mineral processes because the mineral being treated contains sufficient iron as an energy source for the culture and to provide iron(III) to assist the primary and secondary mineral oxidations.

Other *Thiobacillus* cultures have different nutrient requirements and either the proportions of the constituents are changed or some other nutrient is added. For example, the nutrient medium for *Thiobacillus intermedius* includes zinc, copper, manganese and molybdenum salts as sources of these elements which are required in trace amounts. Moderately thermophilic cultures and *Sulfolobus* cultures also have different nutrient requirements which vary with the source of the culture and, in some cases, the type of sulfide mineral to be treated.

Each culture has its special nutrient requirements to allow it to function at its optimum level and each application must be carefully assessed to ensure that the nutrient level is sufficient without being excessive. Accurate analysis of the process water supply and the mineral matrix is necessary to determine the optimum conditions if much trial-and-error testing at the plant stage is to be avoided.

It is normal, in the static methods, for there to be no added nutrients, use being made of those already present in the water and the solid materials. If necessary nutrients could be added to the water supply used in heap, vat and dump leaching operations.

6.2.6 Culture growth

The interdependence of culture growth, particle size distribution and pulp density is dealt with in section 6.2.1 above. The growth rate of the selected bacterial culture is important in establishing and maintaining its level in a process. Most importantly, a selected culture introduced into a process environment must grow more rapidly than, and become dominant over, any indigenous bacterial species if proper control of the process is to be achieved. If the culture does not outgrow the indigenous species the composition of the bacterial culture will alter over time and control of the process may be lost. The growth and dominance of a selected culture can be influenced by nutrients, temperature, acidity and the ability of the culture to adjust to the concentrations of dissolved metals released by the oxidation process. The selected culture can be inadvertently altered if any of these parameters are forced in a particular direction to try to improve the process, sometimes with a negative effect.

The application of bacterial cultures in agitated reactor systems places special emphasis on bacterial growth. A renewable supply of bacteria is required to attach itself to the fresh sulfide mineral as it enters the reactor. This supply can either be produced *in situ* in the initial stages of a series of reactors or generated in a separate reactor and added to the reacting system as required. The growth rate of the bacterial culture under a specific set of conditions must be known to ensure that sufficient bacteria are available to avoid washout (see section 6.2.1, above). If washout does occur the bacterial culture must be re-established before the oxidation can proceed.

In the static methods of bacterial oxidation there is little control over the growth of the culture. It is hoped that the level of oxygen, carbon dioxide, acidity and nutrients in the percolating solution, together with inorganic nutrients in the mineral matrix, are sufficient to maintain the culture until a sufficient fraction of the sulfide mineral has been solubilized to make the process economic.

6.3 BACTERIAL OXIDATION METHODS

Bacterial oxidation can be applied by a number of different methods as indicated, and discussed briefly, in Chapter 2 and in section 6.1. The choice of method for any particular operation depends upon the sulfide minerals present, the concentration of the valuable metal and the location of the treatment operation.

6.3.1 Agitated reactor vessels

Bacterial oxidation of ground mineral slurry can be carried out in aerated agitated vessels, when the value of the metal is sufficient to justify the cost of installing and operating the equipment. The method is suited to higher value metals (see section 2.3) such as gold (liberation from refractory minerals), silver (as an adjunct to gold recovery), cobalt, uranium, nickel, molybdenum, tin (from sulfide concentrate) and possibly copper (if there is sufficient gold or other metals of value present). The reaction vessel can be either (i) a mechanically stirred tank with a means of introducing air into the slurry, or (ii) an air agitated pachuca reactor where the introduction of the air also provides the agitation of the slurry.

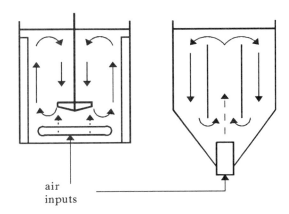

Fig. 6.1 Diagrams of agitated tank (left) and pachuca (right) reactors; air (dotted) and slurry (full) flows are indicated by the arrows

6.3.1.1 Comparison of agitated reactor types

The basic details of the two types of agitated reactors are shown in Fig. 6.1, the flows of slurry and air being indicated by the solid and open arrows respectively. The mechanically stirred reactor is a conventional cylindrical tank fitted with a motor-gearbox and a suitable impeller. Baffles are incorporated into the sides of the tank to enhance mixing and air is injected into the slurry so that it is evenly dispersed by the impeller on the agitation system. A means of controlling the slurry temperature is normally fitted to the inside of the reactor, although the reactor surface is able to dissipate sufficient heat to provide cooling in some situations.

The pachuca reactor is a cylindrical tank with a conical base. A draught tube is sometimes added to improve the flow pattern and the suspension of the mineral particles in the reactor. The air flow through the slurry results in substantial evaporation of water and this aids the cooling of the system. A means of heating can be installed inside the reactor if the slurry temperature needs to be increased. Air is injected into the base of the cone of the pachuca so that it travels up the centre of the reactor generating a flow of slurry within the vessel. The air only mixes with the slurry as it travels upwards and there is little or no entrainment of bubbles in the downward flow. The air lift principle limits the amount of mineral solids per unit volume that can be suspended and it may cause dead zones in the reactor where there is little aeration or agitation of the slurry.

The separation of the aeration and agitation duties in the mechanically stirred reactor allows their separate optimization to suit a particular application or to compensate for the variations that occur in real operating circumstances. The air flow rate is lower than that for a pachuca reactor and there is a reduced incidence of the frothing caused by the presence of residual flotation reagents and by the organic compounds produced by the culture.

The mechanically agitated reactor has obvious advantages over the pachuca reactor and is more versatile. The variations in this type of reactor and its applications are discussed in more detail in the next sections.

6.3.1.2 Types of agitators

Agitation is critical to the bacterial oxidation process. It must be sufficient to suspend the mineral particles to form a slurry, to disperse the air injected into the slurry, to move the slurry around the heat exchangers to maintain the correct operating temperature, to mix the nutrients evenly in the slurry and to remove the reaction products from the oxidizing mineral surfaces.

Mechanically agitated reactors use an electric motor-gearbox unit driving a shaft with one or more impellers. The type of impeller is varied to match the air dispersion and solids suspension duties of a particular process. Impellers may be either high shear or low shear design. Fig. 6.2 shows examples of the two types of impellers which have been used by the authors in laboratory work.

Fig 6.2 A photographs of laboratory impellers;
high shear (left), low shear (right)

The choice of impeller depends on whether the impeller is required to break up the air into fine bubbles or just to disperse fine bubbles produced by a separate aeration system. The former case was thought to require a high shear impeller such as a Rushton turbine (Fig. 6.2 (left)) whereas the dispersion of air bubbles from a separate generator could use a low shear impeller (Fig. 6.2 (right)). Recent developments in impeller technology have resulted in lower shear impellers that are capable of providing sufficient agitation to keep the solids in an evenly mixed suspension and of forming and dispersing the fine air bubbles needed for bacterial oxidation.

The practical implications of the choice of impeller type are that the high shear impellers require up to ten times the power of the low shear variety to achieve the air dispersion and agitation necessary for bacterial oxidation. They suffer from high wear rates and are inefficient at producing an evenly mixed slurry. Another disadvantage of high shear impellers is that they cause considerable heating of the stirred slurry. All of these deficiencies affect the cost of bacterial oxidation and must be translated into positive benefits if the use of high shear impellers is to be justified.

6.3.1.3 Aeration

Aeration is necessary to provide oxygen for bacterial respiration and carbon dioxide for bacterial growth. The finer the bubble size and the longer the air remains as small bubbles, the higher is the transfer rate of oxygen and carbon dioxide into the slurry where they can be used by the bacterial culture. The bacterial oxidation process will be slower than optimum and may even cease if insufficient air is injected into the slurry. The dissolved oxygen level should be maintained at greater than 2 mg L^{-1} to ensure adequate bacterial oxidation rates.

Aeration can be provided by placing air injection points adjacent to the impeller so that fine bubbles are formed as the air passes the impeller tips. An air dispersion pad, consisting usually of a slotted rubber sheet on a supporting frame, is a second common method of generating fine bubbles.

The location of the air injection points with respect to the impeller blades, in those systems where the impeller is used to reduce the bubble size, is variable. One manufacturer advocates an air ring below the impeller so that the air is dispersed as it rises through the slurry. This configuration reduces the risk of the air flooding the impeller and decreasing the slurry flow but loses some of the benefit of the impeller reducing the bubble size. A second manufacturer recommends an air ring slightly above the impeller so that most of the air bubbles are reduced in size by the tips of the blades. This configuration can produce a finer bubble size but there is some risk of flooding the impeller and reducing the slurry flow if the air flow is not matched to the impeller.

The location of air dispersion pads is arbitrary, provided that the fine bubbles generated by the pads are mixed in with the slurry and do not coalesce and rise quickly to the surface of the slurry. The use of air dispersion pads appears to be preferable because the bubbles are formed and their size controlled by the pad. In practice problems associated with wear of the pad and blockage due to iron oxide formation or calcium scaling can offset these advantages. The cost of using pads for air dispersion adds to the capital and operating cost of the reactors and must be justified by improvements in the bacterial oxidation process if aeration is to be provided in this manner.

The theoretical amount of air required per unit time can be calculated from the amount of sulfide mineral reacted in that time if a particular oxidation reaction is assumed to be occurring. The actual quantity used in a continuously operating system would then be determined by the efficiency of transfer of the oxygen and carbon dioxide into the solution so that they can be used by the bacterial culture. This efficiency can be measured experimentally at laboratory and pilot plant scale. Since different minerals undergo bacterial oxidation at different rates and since the majority of sulfide slurries treated are mixtures of minerals, such assumptions are necessarily subject to considerable error. In most sulfide mineral oxidations it is difficult to determine exactly which reactions are occurring and it is only possible to estimate an overall air requirement for the oxidation process.

6.3.1.4 Temperature control

Temperature control of the reactor is important if the bacterial culture is to perform at its optimum level. Low temperatures will result in lower rates of sulfide mineral oxidation and higher temperatures will cause the culture to either become dormant or to become non-viable.

Bacterial oxidation systems which have a net heat generation require cooling to keep the temperature below that which will retard or kill the bacterial culture. This cooling is most economically achieved by the use of heat exchangers in the slurry and water cooling via evaporative cooling towers. Such systems which rely on evaporative cooling can be difficult to operate in areas of high humidity and high temperatures, where there is a shortage of water, and areas where the water contains significant scale-forming minerals, such as calcium and magnesium salts. A second major disadvantage is that a failure in the cooling system may cause the loss of the bacterial culture. If this happens a considerable time will be lost before sufficient culture can be grown from new stock to restart the bacterial oxidation process.

Bacterial oxidation systems that have a net heat requirement are more easily controlled. The process temperature is maintained by circulation of water, usually around 10°C higher than the operating temperature, through heat exchangers in the slurry. These systems benefit from high ambient temperature and humidity, do not require large quantities of water, can use waste low grade heat for compressor or generator cooling systems as a heat source, and are not affected greatly by scale forming minerals because the heating system is a closed circuit and the water used can be given an appropriate softening treatment. Failure of temperature control in these systems results in a fall in process temperature and a loss of bacterial activity. Raising the temperature back to its correct level restores the bacterial oxidation rate to its optimum value within a relatively short time.

The heat in a bacterial oxidation process can be generated by chemical reaction, bacterial reaction, mechanical friction from agitation and from the air injected into the slurry. Heat can be lost from the system through conduction through the reactor walls, evaporation of water by the injected air and in the slurry discharging from the last bacterial oxidation stage. Although mineral oxidation reactions are exothermic only a small percentage of the theoretical exothermicity contributes to the heating of the slurry. This is because the bacterial growth makes use of the majority of the energy available from the oxidation process, the chemical energy of the mineral being used in the synthesis of bacterial cells.

The reactor design can have a major effect on the heat balance and the temperature control of bacterial oxidation in agitated reactors. Selection of a low shear impeller reduces the amount of frictional heat because all the power used in agitation is ultimately converted into heat. Cooling of the injected air can minimize heat input from this source as can the use of compressors rather than air blowers. Use of rubber lined tanks rather than ones constructed from stainless steel reduces heat loss by conduction through the reactor surfaces and covering the reactors reduces evaporative heat losses.

Some chemical reaction is inevitable, with the associated heat production. Addition of sulfuric acid to control the pH value, for instance, generates heat due to the dilution and neutralization of the acid. Likewise, the addition of slaked lime (used in some operations) to increase the pH value generates heat.

The ideal bacterial oxidation process would be heat neutral so that neither heating nor cooling are necessary to maintain the operating temperature. In practice, variation in the amount and type of sulfide mineral entering the process and in the ambient temperature prevent this ideal situation from being achieved. A slight net heat requirement is much the preferred option (compared to any system which requires cooling) as it simplifies temperature control and because failure of the control system is not likely to have catastrophic consequences.

6.3.1.5 Applications

Agitated reactors can be used to treat sulfide material using bacterial oxidation. However, such reactors are normally considered only for the oxidation of high value ores and concentrates such as gold, copper/gold and high grade nickel because of the costs of constructing and operating the processing plant. Fig. 6.3 is a block diagram showing the main features of a typical bacterial oxidation plant. The 2:1:1 layout of the agitators is the optimal arrangement and allows minimal short-circuiting of the feed material.

The associated flow sheet is shown in Fig. 6.4 for the bacterial oxidation treatment of a gold-bearing refractory concentrate. That for the bacterial oxidation of a base metal sulfide concentrate is shown in Fig. 6.5. The flow sheet for the treatment of the waste solution follows the same route as the liquid effluent from the gold processing flow sheet shown in Fig. 6.4.

The construction costs are high because agitated reactor plants must be constructed on properly prepared foundations capable of supporting the weight of the equipment, the plant services and the tank contents. They must be constructed from materials which are able to withstand the aggressive chemical conditions that exist in the bacterial oxidation process. A large number of plant services must be connected to each reactor such as air, water for temperature control, acid or lime to control the acidity, power for mechanical agitation and instrumentation and the electrical circuits for plant control.

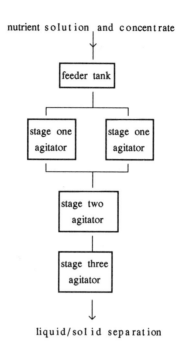

Fig. 6.3 A flow sheet for a typical bacterial oxidation plant

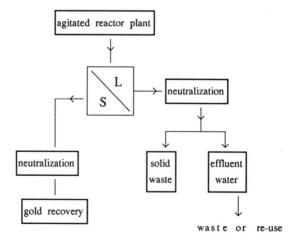

Fig. 6.4 Flow sheet for gold processing by the agitated reactor method

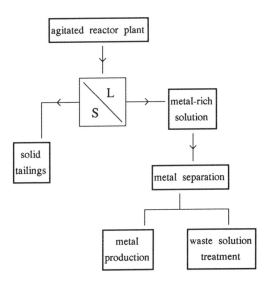

Fig. 6.5 Flow sheet for base metal production by the agitated reactor method

The operating costs are relatively high primarily because of the electrical power consumed in preparing a slurry of finely ground sulfide particles, agitation of the slurry and injection of air into the slurry. In arid areas the quantity of water required in the process can also be a significant cost. The operating costs associated with labour, acid, maintenance and plant control are comparatively low.

6.3.2 Heaps

Heaps are formed by stacking crushed rock into carefully constructed piles on prepared impervious pads which have a sloping base to allow solution to flow by gravity into collection drains and sumps. The base may be rendered impervious by using compacted clay, asphalt, PVC (polyvinyl chloride) sheeting, or concrete lined with HDPE (high density polyethylene) sheeting. The selection of the base is partly determined by the expected duration of the bacterial oxidation treatment and the grade of the ore being treated. The size of the rock pieces can range from several metres down to small clay particles. Normally the largest size of rock is less than 250 mm diameter and the size of particles is spread over the entire range. A distribution of differently sized particles is essential to prevent the flow of solution from the surface through the heap being too rapid. A heap constructed from identically sized particles would allow too great a flow of solution to be practicable.

6.3.2.1 Construction of heaps

The heaps used for the bacterial oxidation of sulfide ores can be constructed by several methods. All heaps have a sloping impervious base to collect the solution as it drains from the base of the crushed rock and each heap must be constructed to be permeable so as to allow an even flow of solution through the crushed rock. An additional consideration for heaps that are used for bacterial oxidation processes is that the heap must have sufficient pore space to allow air to diffuse into the heap to sustain the bacterial culture. The flow sheet for the heap treatment of a gold ore or an agglomerated low grade gold concentrate is similar to that shown in Fig. 6.4 for the agitated reactor method. The only difference is that, in the heap leaching case, the liquid/solid separation occurs by gravity. The flow sheet for a heap leaching operation for the extraction of base metals is similar to that shown in Fig. 6.5 with the exception of any mechanical liquid/solid separation being necessary, the metal-depleted solids remaining in place.

A sloping earth base is essential for ore heaps used for bacterial oxidation, as is the case with those used for direct metal recovery. The base usually slopes diagonally across the rectangular area where the ore is to be stacked. The slope must be steep enough so that the solution draining through the stacked ore collects and flows to the collection drains at the side of the heaps. A slope of three to five degrees is usually sufficient to provide this flow but greater angles can be used.

The sloping base is covered with an impermeable layer. This may be placed directly on the base if it is sufficiently robust such as asphalt, welded HDPE or coated concrete. Slimes from old tailings dams, sand or clay must be placed on the prepared, sloping base before the more common and less expensive PVC membranes are positioned. The placement of these materials prevents the penetration of the membrane by sharp rocks remaining in the surface of the prepared pad, assists with filling minor depressions left in the surface after preparation of the sloping base, and provides a cushioning effect when the ore is placed onto the heap.

The most common methods for placing the crushed rock onto a heap are (i) movable/slewing conveyors, (ii) clam shell buckets and (iii) dumping from haul trucks and pushing into place with a bulldozer. Each construction method has some potential to create preferred solution flow through the heap thereby preventing complete reaction of all the sulfide mineral by the bacterial oxidation process.

(i) Movable/slewing conveyor systems

These consist of a long movable conveyor that transports the crushed ore from the crushing and screening plant to the distribution (slewing) conveyor. The movable conveyor is able to be extended or shortened depending on the distance down the prepared pad that the ore is being placed. The ore is placed first on the end of the pad furthest away from the ore preparation area. The conveyor is progressively shortened as the pad is filled and the heap constructed to its full height. The crushed ore is transferred to the slewing conveyor near the point where it is to be placed on the pad. This conveyor is able to sweep through a broad arc so that the ore can be distributed across the width of the pad up to the final height of the heap. Several passes of the slewing conveyor are normally used to build the heap up to its final height. This minimizes the grading of the ore into segregated sizes which would encourage preferred solution flow.

The major benefit of this method of heap construction is that large quantities of ore can be moved efficiently and placed onto the heap in a uniform manner. The slewing conveyor can be remotely controlled and limited automation of the ore stacking procedure is possible. This control over the stacking of the crushed ore reduces the incidence of preferred solution flow through the heap and creates conditions for more consistent bacterial oxidation activity in the heap because of the relative uniformity.

One of the problems of using this method of heap construction is the need to move sections of the conveyor continually as the heap construction proceeds. The conveyor equipment must be sufficiently mobile to be moved easily but this adds to the cost. It is then unproductive until the next pad is filled with ore. Much of the equipment is under utilized which adds to the cost of the overall process. A second problem is that it may not be possible to irrigate the heap and recover the solution until the entire pad is filled with ore. It is not possible to move the conveyor sections over the pad lining material without causing damage, at least in the cases of the less expensive lining materials. If the ore preparation area and the solution collection system are at the same end of the heaps and it is desirable to start treating the ore as each section of the heap is constructed then a temporary side drain and a bund wall system must be prepared as each new section of heap is added to that already being treated. Each section of temporary drain or bund wall constructed to retain solution adds to the cost of treatment. If the collection ponds and solution treatment areas are on the lower end of the prepared pads and the ore preparation plant at the higher end so the conveyor can stack while the recently place ore is irrigated, there is more cost in constructing and operating the process. Power must be distributed over a greater area and process supervision is more complicated because of the separation of the equipment that needs closest monitoring.

(ii) Clam shell buckets

This method consists of a tracked vehicle able to rotate on the tracked base with a boom from which a two-piece clam shell bucket is suspended by cables. The tracked base allows horizontal movement and rotation to the position where ore is to be added to the heap under construction. The cables allow the bucket to be raised, lowered, opened and closed. The boom allows for elevation and reach as the heap is constructed. The two - piece bucket opens and then closes to scoop up ore. The bucket is opened when the bucket is correctly positioned to release the ore into the depression between ore already placed on the heap.

The mobility of the clam shell bucket allows the ore to be added to heap from the most convenient position. Normally the ore is added at the side of the heap higher up the sloping leach pads. This allows the membrane to be extended just behind the movement of the clam shell bucket so that the reach of the bucket allows it to always be positioned at the edge of the membrane to place the ore. The pipes can be placed and the solution applied to the heaps immediately because the solution flows down the gradient to the collection drains, away from the stacking equipment.

Placement of ore onto heaps using clam shell buckets can achieve relative uniformity in the constructed heaps provided sufficient care is taken by the operator. The clam shell bucket places crushed ore in small cones. New cones of ore must be carefully placed into depressions between existing ore to prevent segregation and preferred solution flow through the heap.

Problems with using clam shell buckets to build constructed heaps for bacterial oxidation are the limitations on the amount of ore that can be stacked in a given time and the reliance on the operator to prevent segregation of the ore. The quantity of ore that can be stacked by the clam shell bucket and its operator is dictated by the size of the bucket, the proximity of the ore to the point where it is added to the stack, and the skill of the operator in stacking the ore. The size of the bucket is determined by the clam shell bucket equipment selected for the stacking process. The ore is normally delivered immediately adjacent to the stacking unit by trucks which must have access to each new position where ore is to be added to the heap. Each time the clam shell bucket moves to a new position, the trucks must create a new pile for the bucket to stack. The limitation on the rate of stacking is normally fixed by the ability of the trucks to keep up the supply of crushed ore from the ore preparation plant. The stacking process is dependent upon the experience of the clam shell bucket operator for the placement of ore to prevent segregation of the ore and to maintain the quantity ore being added to the heap.

Thus using a clam shell bucket can provide an effective method for constructing heaps for bacterial oxidation but it is dependent on correct equipment selection and an experienced operator. This method does have the advantage that the ore recently added to the heap can be irrigated as soon as the section of the heap has been finished. There are limitations on the quantity of the ore that can be stacked which are imposed by both the selection of equipment and the operator. This method is more labour intensive than the moving/slewing conveyor method of ore placement.

(iii) Haul trucks plus bulldozer.

A combination of haul trucks and a bulldozer suitable for final placement of the ore into heaps can be used to construct heaps for bacterial oxidation. The ore is taken from the ore preparation plant and dumped directly onto the membrane. Ore is pushed into place by a suitable bulldozer, usually one with low ground pressure to prevent compaction.

This method of constructing a heap for bacterial oxidation is very susceptible to producing segregation of the ore. Ore segregates as it dumps from the haul trucks so that the finer material moves to the bottom of each pile and the coarser material remains on the top and upward sides of the pile dumped by the truck. Much of the ore is not moved from the position where it was dumped so that the selected size distribution of ore remains. Additional segregation occurs as the bulldozer pushes the ore up to form the heap. The larger sizes separate to the top and the finer particles flow through the spaces to the bottom of the ore being moved.

The operator of the bulldozer must form the heap to its final height. Often the bulldozer must traverse parts of the pad where ore has already been placed causing compaction of the ore.

The porosity of the heap is substantially reduced by the segregation of the ore, by the filling of voids by fine particles as the ore is pushed up to its final position and by compaction under the weight of the bulldozer. The reduced porosity restricts the rate of application of solution to the heap and may prevent some areas of the heap from being wetted because of an impermeable, compacted layer within the heap. The separation of the fines from the coarse pieces of rock results in preferred solution flow within the heap and causes an uneven reaction rate so that some parts are left unreacted when most of the sulfide mineral has been removed.

The rate at which ore can be added to the heap being constructed is determined by the amount the trucks can deliver, the size of the equipment and how quickly the bulldozer operator can push ore to the final height. This method of construction is faster than the other two described but produces a less uniform heap. The skill of the bulldozer operator can have a large influence on the segregation and the solution percolation through the ore but it is extremely difficult not to have layers of finer particles and compaction in some parts of the heap.

Solution application to the ore is achieved by pumps feeding a networks of pipes fitted with a means of spreading solution over the entire surface of the ore. The solution should be applied evenly to produce comparable reaction in all parts of the ore in the heap.

Many methods of distributing the solution have been tried. They involve the use of either drippers, 'wigglers' or 'wobblers'. Small drippers or emitters provide a slow flow of solution onto ore immediately adjacent to these fittings and rely on regular placement and careful heap construction to wet the areas of ore between these fittings and the rows of pipes. Wigglers are short lengths of flexible tube and distribute solution by using the irregular whipping action caused by solution flow under pressure from the end of the tube. These wigglers are capable of a wide distribution of solution which is determined by the diameter of the tubing and the pressure at the discharge. The wigglers are fitted along the length of the pipes and the pipes are spaced so that the areas covered by each wiggler overlap to give total coverage of the heap surface. Wobblers are a type of oscillating sprinkler which randomly distribute solution over a 360° radius which is determined by the size of wobbler and the pressure of the solution being discharged. The wobblers are placed at regular intervals along the pipes and the pipes are spaced so that the radii of the solution distribution overlap.

Drippers and wigglers both suffer from blocking by precipitation of the dissolved solids when the solution contacts with the air. The higher flow rates through the wobblers dislodge most of the precipitate and prevent the build up blocking the discharge. Wigglers and wobblers discharge the solution into the air so that a high percentage of the solution is lost through evaporation in dry, windy areas. Discharge of the solution into the air aids the dissolution of oxygen and carbon dioxide and maximizes their application to the heap. Drippers, which apply the solution directly onto the surface of the heap, are more suited to regions where the evaporation rate is high.

6.3.2.2 Control of bacterial oxidation in heaps

Some operating parameters cannot be controlled in a heap leaching system and the control of others is limited compared with agitated systems. The parameters that cannot be kept in a specified range include temperature and culture growth. The acidity, the nutrients, aeration and the quantity of sulfide mineral can be controlled to a limited extent, either by the manner in which the heap is constructed, by the irrigation of the heap with the bacterial oxidation solution or the adjustment of the solution before application to the heap.

The temperature in the heap cannot be kept within a specified range. The ambient temperature, the size of the heap, the quantity of sulfide mineral and the amount of the chemical reaction will all influence the temperature within the heap. High internal temperatures have been observed in some situations where reactive sulfide minerals have been present in large quantities.

Culture growth cannot be influenced within the heap other than by trying to maintain the acidity, the supply of nutrients and the transfer of oxygen and carbon dioxide by convection, diffusion and solution flow.

The acidity within the heap can be controlled on a macroscopic scale by adjusting the pH value of the solution being applied to the surface of the heap. In the bulk of the heap, the acidity is determined by the uniformity of solution flow through the stacked rock and by the distribution of acid consuming minerals within the heap. Localized variations are common but most bacterial cultures operate over a wide range of acidity so that the heap system can normally be maintained within a range where the bacterial oxidation process can be sustained.

The nutrients required to sustain bacterial oxidation can come from either the ore (rock) or the solution percolating through the heap material. Chemical nutrients for bacterial oxidation can be added to the solution being applied to the surface of the heap to make up any deficiency of nutrients in the ore. Difficulties can be experienced where nutrients are higher than optimum as reduction in nutrient levels in ore are not economically practical in most cases.

Aeration within the heap is achieved by convection and diffusion of air through the pore spaces in the heap and by transfer in the solution being applied to the heap (Lungdren and Malouf, 1983). The construction of the heap determines the extent of transfer of oxygen and carbon dioxide directly from the air and to a lesser extent the transfer of these gases by the solution. Sufficient porosity should be built into the heap during the construction to allow air to diffuse to the centre of the heap. This pore space also allows free movement and relatively uniform distribution of solution which is the second means of transporting the gases into the heap. The application rate of solution to the surface of the heap can be used to increase the gases transferred by solution flow. However, the pore spaces are filled by solution, reducing the convective and diffusion transfer, as the amount of solution flowing through the heap increases. The method of application of solution influences the amount of oxygen introduced in to the heap per unit volume of solution. Drippers allow little contact of the solution with the air to absorb oxygen whereas wigglers and wobblers spray droplets through the air allowing larger quantities of oxygen to be absorbed into the bacterial oxidation solution being applied to the heap.

In some cases where the sulfide level is high or the mineral is particularly reactive, positive aeration has been suggested. Pipes could be laid underneath the ore and compressed air pumped into the base of the heap to increase the air transfer. This form of aeration has the added cost of the piping, the compressor and the power to operate the system and would have to be justified by the faster extraction of valuable metal or by producing a higher recovery of metal.

The quantity and distribution of the sulfide minerals in the heap are primarily dictated by the ore being treated but can be managed to a limited extent by the stacking process when the heap is being constructed. An ore stacking system that allows the operator to distribute the sulfide mineral in the heap evenly as it is delivered from the ore preparation plant will produce a more regular bacterial oxidation reaction within the heap. The air and nutrient requirements, the consumption or production of acid and the variation of temperature are all more uniform if the sulfide mineral and, thus, the bacterial activity is more evenly distributed in the heap.

Most heap systems contain iron minerals that release soluble iron during the bacterial oxidation process. If this dissolved iron is oxidized to the iron(III) state within the heap and the acidity is too low, then precipitates of hydrated iron oxides can form. These precipitates are capable of coating the sulfide minerals or blocking the pore spaces and reducing the permeability of the heap to both solution and air. The net result is a lower rate of oxidation or a reduced amount of oxidation of the sulfide minerals. Both phenomena reduce the recovery of the valuable metal, at least in the short term, thereby reducing the viability of the bacterial oxidation process.

Bacterial oxidation of ores to recover metals that are insoluble in acid sulfate solutions require additional treatment. Examples of metals that fall into this category are gold, silver and lead. Washing to remove iron and acid followed by neutralization with an alkali is necessary before silver and gold can be extracted using conventional cyanide leaching. Incorrect washing or neutralization results in lower metal recovery and higher cyanide consumption. When lead sulfide undergoes bacterial oxidation it is converted to insoluble lead sulfate. This can be leached out by treatment with sodium thiosulfate solution. Removal of excess acid before extraction of the lead by thiosulfate leaching is essential to prevent disproportionation of the thiosulfate ions to sulfite ions and elemental sulfur.

The acidity must also be controlled to reduce the amount of oxidation of the thiosulfate by the bacterial culture within the heap.

6.3.3 Dumps

Bacterial oxidation can and does occur in waste and low grade ore dumps, often without any encouragement and sometimes to the detriment of the environment. These dumps are constructed from the broken rock which is mined to expose the more valuable ore, but which is too low in metal value to be treated in agitated reactors or prepared heaps or vats.

Waste and low grade ore dumps usually contain small quantities of sulfide minerals which are dispersed throughout the waste rock as well as indigenous sulfur oxidizing bacteria. The size range of the rock can vary from several metres in diameter to fine clay particles. The flow sheet for the base metal solubilization in a dump operation is similar to that shown in Fig. 6.5 for the agitated tank method. There is no mechanical liquid/solid separation as the separation occurs by gravity. The metal-depleted solid remains in place.

6.3.3.1 Construction and location of dumps

Bacterial activity does take place in dumps which were not constructed for the purpose. However, now that bacterial oxidation of sulfidic mineral content of waste rock is well established, it is advantageous to construct dumps so that the process operates optimally. Bacterial oxidation dumps are constructed on an impervious base to contain the bacterial oxidation solution and are often located on the side of a slope or in a valley so that this solution flows through by gravity and collects in a sump at the base. The broken rock, with no further comminution, is dumped from trucks or a moving conveyor with no particular care being taken in the construction of the dump except to ensure the dumped material is not compacted to the point where solution cannot percolate. The solution used to promote bacterial oxidation is applied by a system of pipes and sprinklers.

6.3.3.2 Control of bacterial oxidation in dumps

The nature of the dump places restrictions on the parameters that can be used to control the bacterial oxidation process. Parameters such as temperature and aeration are dictated by the construction of the dump and the nature of the sulfide minerals in the rock. Acidity, solution flow and nutrients can be controlled at the point of application at the surface of the dump but no control within the dump is possible. The bacterial culture that will exist within the dump will be determined by the local environment within the dump and the composition of the culture is likely to vary at different locations.

The acidity of the bacterial oxidation solution applied to the surface of the dump can be adjusted so that the pH value of the solution within the dump is favorable for bacterial growth and oxidation of sulfide. The upper limit for acidity is dictated by the greatest concentration that can be tolerated by the bacterial culture being used. The lower limit is controlled by monitoring the pH of the discharge solution so that bacterial oxidation is maintained and iron oxide precipitates do not form and reduce the solution flow through the dump.

The solution flow through the dump provides some aeration, assists the transfer of heat through the dump, disperses the nutrients to the bacteria and maintains the acidity on a macroscopic level. The maximum solution flow is dictated by the construction of the dump, the porosity of the rock and the particle size distribution, particularly the fine clay particles.

The nutrients are supplied to the bacterial culture to increase the reaction rate. Only those nutrients not already available in the rock or the water supply are added and in many cases no addition is necessary.

The bacterial oxidation process in sulfidic dumps normally uses the indigenous bacterial species present but another culture can be introduced if the indigenous one is not capable of providing a reasonable rate of reaction. The introduced culture must be able to outgrow the indigenous species under the conditions within the dump if any improvement in reaction rate is to result. The selection of bacterial culture is made by control of those parameters that provide conditions closest to optimum for the culture of interest. However, close control of the environment within the dump is impossible and problems of overheating and preferential solution flow can arise. The introduction of a bacterial culture into a dump situation should be carefully considered as it may not be successful.

6.3.4 Vats

Vat leaching refers to the method of treatment where the sulfide mineral is immersed in the solution for all or part of the treatment process. Most often this form of treatment is used to treat old tailings which have previously been separated into 'sands' and 'slimes'. The 'sands' fraction comprises the coarser material that may not have been fully reacted in previous treatment, typically when the metal price was controlled or very low and only high grade ore was treated. These tailings are now often higher in metal value than some of the oxide or more modern ore deposits being treated. Only the 'sands' fraction of the old tailings can be treated by this method of bacterial oxidation. The flow sheet for a vat operation is similar to that of heap leaching (see section 6.3.2.1 and Figs 6.4 and 6.5).

Modern vats are mostly constructed by building a bund wall, lining the dammed area with an impervious layer and providing a means of solution circulation. The base of the vat is normally sloped to one corner to facilitate solution drainage and recycling. PVC liners are most common because of their low cost and their ability to withstand chemical attack. The liner can be very much thinner than those for heap leaching because these liners only have relatively fine material dumped onto the prepared base and do not have to withstand the impact of larger rocks.

A vat normally has a means of solution distribution over the surface of the material being treated, with the majority of the solution being applied at the end of the vat furthest from the solution recovery point. In the crudest form, solution is pumped into a sump at the point furthest from the solution recovery area. This flow pattern allows the solution to travel the greatest distance through the ore or tailings being treated and gives more opportunity for bacterial oxidation reaction.

The coarse nature of the material being treated allows the bacterial oxidation solution to flow through the bed, thereby distributing nutrients and oxygen, maintaining the acidity level and removing reaction products from the system. If the material is too fine then adequate percolation of solution is not possible and the bacterial reaction is retarded.

Aeration of the bed can be assisted by completely draining the vat periodically to allow diffusion of air through the pore spaces. The additional oxygen provided by this wetting and draining cycle has the potential to cause a substantial increase in the rate of the bacterial oxidation reaction which would otherwise be slow in a flooded and relatively deaerated bed.

Vat leaching requires substantial amounts of water because of the need to flood the bed of material being treated. This requirement limits this treatment method to areas with an adequate water supply. This treatment method is also unsuited to very high rainfall areas because the vat would continually overfill due to the precipitation making control of the bacterial oxidation solution conditions extremely difficult or impossible.

6.3.5 *In situ* treatment of ore bodies

In situ, or in-place, treatment refers to the process where broken ore is treated without mining to remove the rock to a treatment process on the surface. This process relies on fracturing the ore and producing sufficient voids and porosity to allow free and relatively uniform solution flow. It also relies on the ore body being sealed by a natural impervious layer to prevent the valuable metal dissolved in the solution from draining into the surrounding rock and being lost. The method is not suitable for an ore body which is highly acid consuming. These constraints make *in situ* leaching a relatively rare form of metals recovery, by bacterial oxidation or chemical leaching. The flow sheet for the base metal solubilization in an *in situ* operation is the same as that following the metal-rich solution shown for base metal production in Fig. 6.5.

The most suitable sites for the application of *in situ* leaching are old mine workings which already contain significant voids in the form of shafts and stopes. When the remaining ore is blasted to fracture the rock, there is ample pore space remaining for solution flow. In fresh rock, where previous mining has not occurred, some of the ore (up to twenty percent) must be removed to allow sufficient porosity for *in situ* leaching to occur after the rock is fractured.

The need for an impervious layer to prevent solution loss restricts this method of treatment to mining areas that are not heavily fractured or faulted, where the ground water levels are low and to areas free from seismic activity. The threat of metal laden solution or solution containing toxic chemicals escaping into the ground water prevents this method of treatment being applied.

The solution necessary for bacterial oxidation is distributed through the broken ore either through a piping system in the case of an old mine workings or by a series of injection wells in the case of a fresh ore body. The solution flows down through the broken rock to a sump or a series of collection wells where it is pumped to the surface for treatment before being returned underground.

The nutrients necessary for bacterial oxidation culture are normally present in the rock and the local water supply because the bacterial culture is normally indigenous to the ore being treated. If supplementary nutrients are required, then these chemicals are added before the solution is injected into the ore.

Oxygen is supplied by the flow of solution through the ore or by diffusion and convection in those areas that have adequate air flow. The oxygen supply can be a limiting factor in bacterial oxidation using an *in situ* treatment method, particularly in ore bodies where the sulfide minerals are reactive and quickly consume the oxygen in the solution. Positive aeration of the solution before pumping into the broken rock can assist the oxygen transfer but must be assessed to insure that the additional cost is justified.

Control of the acidity is difficult for *in situ* bacterial oxidation. The presence of acid consuming minerals can cause a rise in pH resulting in precipitation of hydrated iron oxides and plugging of parts of the ore body. Reactive sulfide minerals can cause excessive, localized acid generation and may lead to reduced bacterial activity until the equilibrium is re-established. The only opportunity to adjust the acidity of the bacterial oxidation solution is when the solution is returned to the surface, either before metal recovery or injection back into the ore body. There is no opportunity for control of temperature in *in situ* bacterial oxidation treatment. If substantial chemical reaction causes a rise in temperature or if the natural rock temperatures are high, then the bacterial oxidation reaction may be retarded until the conditions fall within the range of operation of the bacterial culture present in the ore.

6.4 EXAMPLES OF PROCESSING METHODS

This section is devoted to a selection of published descriptions of the processing of minerals by bacterial methods.

6.4.1 Tank leaching

All tank leaching operations to date have been restricted to the processing of refractory gold-bearing materials. Hackl *et al.* (1989) describe the plant and its economics for the processing of ten tons per day of a concentrate at the Salmita mine.

Spencer *et al.* (1991) have described the treatment of a concentrate, containing seventeen percent of arsenic, in a large pilot plant with a throughput of one tonne per day. The treatments of refractory gold-bearing concentrates containing appreciable amounts of copper, nickel, zinc and silver were described by Spencer *et al.* (1988).

A very detailed account (Marchant *et al*. 1989) is available of a feasibility project study for the treatment of a refractory carbonaceous sulfide concentrate at Austin, Nevada. There are descriptions of the results of testing at pilot plant level which lead to design parameters for large scale plants for the treatment of 20-40 tonnes of sulfide concentrate per day.

Van Aswegen *et al*. (1988) give an account of the design and operation of the plant at the Fairview mine which was treating a concentrate at the rate of twelve tons per day. The plant has since been expanded to treat the total production of concentrate of the mine of forty tons per day (Mining Magazine, 1991).

6.4.2 Heap leaching

An extensive description of the heap leaching of a porphyritic copper sulfide deposit is given by Durand and Walqui (1990). The heaps cover an area of one million square metres and are sprinkled with an acidic solution. The sulfide ore requires treatment for between ninety and one hundred and twenty days. The solution effluent is subjected to liquid ion exchange using the LIX 864 reagent and, after stripping, the copper is produced by an electrowinning stage.

There are no published reports to date of heap leaching operations which are operated to liberate gold from its refractory encapsulating minerals.

6.4.3 Vat leaching

This has been carried out at the Inspiration mine in Arizona since the 1930s and is described by Aldrich and Scott (1933). Although the intervention of microbes was not considered it is highly likely that the leaching reactions are bacterially catalysed.

6.4.4 Dump leaching

This method of producing copper from low grade ore or waste dumps is relatively widely used and well documented. Rossi's book (1990) contains a brief but comprehensive summary of all the published descriptions of dump leaching operations.

Vassilev and Groudev (1990) describe a Bulgarian operation for the leaching of copper from a thirty million ton dump of waste low grade ore.

6.4.5 *In situ* leaching

Burton *et al*. (1990) discuss geotechnical factors in the underground leaching of chalcopyrite ores in the Avoca mine in Ireland. They conclude that the reaction rate is limited by the supply of oxygen to the ore. They make reference to an earlier paper (1983) in which they suggested that the primary leaching agent at the chalcopyrite surface is iron(III) and that the function of the bacteria was to regenerate it. They suggest that, if possible, bacterial activity should be maintained throughout the ore body. It is possible that in all the static leaching methods the majority of the primary oxidation in the top portion of the material is due to the direct participation of oxygen rather than iron(III). The iron(III) is much better able to diffuse through the material and it could be that it is the primary oxidant further down the ore pile. If there is a lack of oxygen in heaps and dumps this should result in the iron in the effluent solution being mainly in the (II) state. An extensive description of the *in situ* leaching of copper at the Cerro de Pasco mine in Peru is given by Delgado (1990).

6.5 REFERENCES

Aldrich, H.W. & Scott, W.G. (1933) *Trans. A.I.M.E.*, **106**, 650.

Brierley, J.A. & Brierley, C.L. (1986) *Thermophiles: General, Molecular and Applied Microbiology.* John Wiley and Sons Inc.

Burton, C.J., Heffernan, J. & Thorne, B. (1983) *Final Technical Report - Avoca In Situ Leach Project D6X11 European Commission, Brussels, Contract No. M.S.M. 112 EIR (H) 1983.*

Burton, C.J., Heffernan, J. & Thorne, B. (1990) *Proceedings of an International Seminar on Dump and Underground Bacterial Leaching of Metals from Ores, Leningrad, 1987.* Karavaiko, G.I., Rossi, G. & Avakyan, Z.A. (eds) Centre for International Projects, USSR State Committee for Environmental Protection, Moscow, p.144.

Delgado, O. (1990) *ibid.*, p.164.

Durand, O. & Walqui, H. (1990) *ibid.*, p.187.

Hackl, R.P., Wright, F.R. and Gormely, L.S. *Biohydrometallurgy - 89, Jackson Hole, 1989.* Salley, J., McReady, R.G.L & Wichlacz, P.L. (eds) p.533.

Kuenen, J.G. & Tuovinen, O.H. (1981) *The Prokaryotes*, Vol. 1, Starr, M.P., Stolp, H., Trueper, H.G., Balons, A. & Schlegel, H.G. (eds) Springer Verlag, Ch. 81.

Lungdren, D.G. & Malouf, E.E. (1983) *Advances in Biotechnological Processes.* Alan R. Liss Inc, NY. p.223-249.

Marchant, P.B., Lawrence, R.W., Chapman, J.T., Brooks, W. & Kuipers, J. (1989) *21st Annual Meeting of the Canadian Mineral Processors, Ottawa, Ontario, Canada, 1989*, p.1.

Mining Magazine **164**, 40, (1991).

Rossi, G. (1991) *Biohydrometallurgy.* McGraw-Hill, p.418-430.

Silverman, M.P. & Lundgren, D.G. (1959) *J. Bacteriol.*, **77**, 642.

Spencer, P.A., Budden, J.R. & Barrett, J. *Trans. Instn Min. Metall. (Sept C: Mineral Process. Extr. Metall)*, **100**, C21, (1991).

Spencer, P.A., Budden, J.R. & Rhodes, M.K. (1988) *Minerals Engng* 4 (7-11), 1143.

van Aswegen, P.C., Haines, A.K. & Marais, H.J. (1988) *Randol Perth International Gold Conference, 1988*, p.144.

Vassilev, D.V. & Groudev, S.N. (1990) *Proceedings of an International Seminar on Dump and Underground Bacterial Leaching of Metals from Ores, Leningrad, 1987.*

Karavaiko, G.I., Rossi, G. & Avakyan, Z.A. (eds) Centre for International Projects, USSR State Committee for Environmental Protection, Moscow, p.202.

7

Product and effluent treatment

7.1 PRODUCTS OF BACTERIAL OXIDATION REACTIONS

The final products of the bacterial oxidation treatment of an ore or mineral concentrate consist of an aqueous solution and the residue of the treated solid. The metal values may reside in one or other or both of these phases. In general there are three cases to be considered.

(i) Class I processes for the treatment of refractory gold/silver-bearing minerals where the metal value is liberated but remains in the solid.

(ii) Class II/III solubilizations of metals from their sulfides, arsenosulfides, arsenides, oxides or carbonates (which include cobalt, uranium, nickel, molybdenum, copper and zinc), where the metal values reside in the aqueous solution.

(iii) Class II oxidations of silver, tin, antimony and lead minerals where the metal values exist mainly in the form of insoluble sulfates.

The three cases require different metal recovery procedures. In all three cases the aqueous effluents, after the removal of the metal values (if there were any), contain iron(III), sulfuric acid and possibly arsenic(V). The treatment of these effluent solutions forms the main subject of this chapter. In practice each potentially treatable substrate would have its own flow-sheet based on the results of appropriate amenability testing.

7.2 RECOVERY OF METAL VALUES

7.2.1 Recovery of liberated gold

The solids from an agitated reactor or the material remaining in the solid state after a heap leaching of gold-bearing refractory minerals contain the liberated elemental metal. They might, in addition, contain elemental silver. If an agitated reactor method is used the oxidized slurry is subjected to solid-liquid separation. With heap leaching such separation occurs as part of the heap operation. In both cases the gold (and/or silver) reside in the solid phase which is in a highly acidic state. Before the oxidized solid residue can proceed to a conventional cyanidation plant (or be treated with sodium cyanide solution in the case of a heap operation) it is subjected to washing and neutralization stages. The aqueous solutions from both tank reactor and heap leaching methods, including the washing solutions, contain iron(III), sulfur(VI) (sulfuric acid) and possibly arsenic(V). Even though gold is the main metal value present there could be quantities of other metals, present in the original concentrate (or ore), which would be solubilized in the bacterial liberation process. The neutralization of the solids can be avoided if, instead of cyanidation, a solution of thiourea and hydrogen peroxide is used to dissolve the gold content. This technology is being used on a minor scale, is very effective, and should be considered for new plants where CIP facilities are not already in place.

7.2.2 Recovery of solubilized metal values

Where the metal values of a mineral substrate are solubilized by the bacterial oxidation process the treatment of the aqueous solution, after its separation from the spent solid material, depends upon the particular metal (or metals) it contains. Examples of ionic species which could be present in the aqueous solution are Co^{2+}, UO_2^{2+}, Ni^{2+}, MoO_2^{2+}, Cu^{2+} and Zn^{2+}. In addition the solution would contain iron(III), sulfur(VI) and possibly arsenic(V). Theoretically it is possible to separate the metal values from the solution by a controlled neutralization with lime. This would remove the iron(III), sulfur(VI) and arsenic(V) as a precipitated mixture of gypsum, iron(III) oxohydroxide and iron(III) arsenate. There are problems of control of lime or limestone addition so that the pH value of the solution does not exceed five. Above this value the solubilized metals would be precipitated as their hydroxides and the co-precipitation of the metal values on the waste precipitate would lead to an inefficient separation. Such problems may be avoided by extracting the metal values selectively at low values of pH. In the case of copper solutions the use of scrap iron to produce a copper metal cement is well proven. Nickel and cobalt could be extracted from solution by electrowinning. Other methods which are available are solvent extraction (used in the case of uranium solubilization) or the use of highly selective liquid ion exchangers (LIX reagents). Metal-rich (loaded) LIX reagent solutions are washed (scrubbed) to remove impurities (e.g. iron(III)) and then stripped to give a pure concentrated aqueous solution of the selected metal. The metal value can then be recovered by electrowinning.

After any extraction process the liquid effluent containing the iron(III), sulfur(VI) and arsenic(V) can then be neutralized with no loss of metal values.

7.2.3 Recovery of metal values from solid residues

If sulfidic minerals of silver, tin, antimony or lead are subjected to bacterial oxidation the metals remain in the solid phase as insoluble sulfates. Extracting the metal values from the solid, after separating it from the aqueous phase, would involve its treatment by a solution of either ammonia (to dissolve any silver) or sodium hydroxide (to dissolve any lead, tin or antimony). Silver would be solubilized as the ammine complex, $[Ag(NH_3)_2]^+$. Lead(II), tin(IV) and antimony(III) and (V) would be solubilized as the anions PbO_2^{2-} (plumbite), $Sn(OH)_6^{2-}$ (stannate), SbO_2^- (antimonite) and $Sb(OH)_6^-$ (antimonate), respectively. Alternatively, lead sulfate may be removed from the solid material by using a solution of sodium thiosulfate, $Na_2S_2O_3$, as the complex ion bis-thiosulfatoplumbate(II) ion, $[Pb(S_2O_3)_2]^{2-}$.

7.3 CLEAN-UP OF AQUEOUS EFFLUENTS

The solution effluent from agitated reactors for the bacterial oxidation of pyrite contains iron(III) and sulfur(VI) at concentrations of around 0.3 M (17 g L^{-1}) and 0.6 M (20 g L^{-1}) respectively. That from the bacterial oxidation of arsenopyrite contains, in addition to the iron(III) and sulfur(VI), concentrations of up to 0.27 M (20 g L^{-1}) of solubilized arsenic(V). Solution effluents from the static methods of bacterially oxidizing minerals contain considerably lower concentrations of iron(III), sulfur(VI) and arsenic(V) than those from the stirred tank reactor method. Independently of the methods of bacterial oxidation of a mineral, the solution effluents share the same problems. After the removal of the solubilized metals of value it is important that the iron(III) and arsenic(V) should be economically and easily disposable in a stable state which complies with any local environmental protection regulations. The long term stability of the waste materials is also a matter which requires serious consideration. The effluent water from the clean-up process should conform with the local regulations concerning the allowable minimum levels of iron and arsenic. It is probable that much of the effluent water would be re-used in the bacterial oxidation plant, especially in areas where the water supply is either scarce and/or of poor quality.

Lime (strictly slaked lime, $Ca(OH)_2$) is usually used to neutralize the solution phase and causes the precipitation of calcium sulfate (gypsum) and a mixture of solids containing iron(III), arsenate and sulfate ions and some unreacted lime. The same products are produced if limestone (calcium carbonate) is used, although the reaction rate is lower than in the case of lime usage. The problem, in either case, is complicated by the initial products being those which are produced rapidly (by kinetically controlled processes), the thermodynamically more stable products being formed, if ever, after some years. Subsequent treatment of the stored material is important and should be controlled to minimize any release of arsenic(V). The presence of water and carbon dioxide are of significant influence in determining the identity of the eventual compounds formed and, in consequence, the stability of the iron and arsenic contents.

There are three approaches to the problems of effluent treatment and control. The first is the mineralogical approach which seeks indications of stable compounds from known minerals with their presumed considerable stability. The second is to model the neutralization behaviour of any effluent solution on the basis of known formation constants and solubility product data. The third is the experimental approach where solutions of known composition are neutralized, with solutions (of sodium or potassium hydroxide) or slaked lime (calcium hydroxide) slurry, the precipitated compounds being identified. The three approaches are the subjects of the discussions in the following sections.

7.4 MINERALOGICAL AND THERMODYNAMIC DATA

This section consists of a discussion of possible model compounds for the precipitation of iron(III) and arsenic(V) in the presence of sulfur(VI) from an examination of known minerals containing such oxidation states of those elements. In addition the available data which indicate which are the thermodynamically most stable compounds for any given solution, after suitable neutralization treatment, are considered.

Although calcium hydroxide is commonly used for the stabilization of iron(III) it is useful to consider the differences between the use of sodium (or potassium) hydroxide and calcium hydroxide as precipitating agents for the treatment of solutions containing iron(III), sulfur(VI) and arsenic(V). There are three aspects of such a consideration. One is the nature of the compounds initially produced, the second is the nature of the ultimately produced, and therefore thermodynamically most stable, compounds and the third consists of the conditions which control the transformation of the initially produced compounds to those which are of ultimate thermodynamic stability.

7.4.1 Compounds containing iron(III) and/or sulfur(VI)

The compounds that may be produced by using either sodium (or potassium) hydroxide or calcium hydroxide (or calcium carbonate) as precipitants are dealt with separately.

7.4.1.1 Sodium (or potassium hydroxide) as precipitant

When hydroxide ion is used to raise the pH value of the effluent solution, from its initial value of around unity to the value where precipitation occurs, there are various possible compounds which may be formed. These include basic iron(III) sulfates (including jarosites) and various formulations of amorphous so-called iron(III) hydroxide, goethite (α-FeO(OH)), akaganeite (β-FeO(OH)), lepidocrocite (γ-FeO(OH)) and hematite (α-Fe$_2$O$_3$).

(i) Basic iron(III) sulfates.

The naturally occurring basic iron(III) sulfates are monoclinic butlerite and orthorhombic parabutlerite, both having the formula, $FeSO_4OH.2H_2O$, amaranite, $FeSO_4OH.3H_2O$, fibroferrite, $FeSO_4OH.5H_2O$, and hohmannite, $Fe_2(SO_4)_2(OH)_2.7H_2O$. It is feasible that the neutralization of the $[FeSO_4]^+$, $[FeHSO_4]^{2+}$ and $[Fe(SO_4)_2]^-$ complexes (see section 4.3.5.3.) could cause their incorporation into precipitates which would be formulated as basic sulfates. It is, however, unlikely that any such compounds would be precipitated. They are fairly soluble in water and may represent formulations of the precursors of the eventually precipitated solids. Their formulae may be regarded as 1:1 stoichiometric mixtures of iron(III) sulfate and iron(III) oxohydroxide, with varying amounts of hydration. The only basic sulfates which are insoluble in acidic solutions are the jarosites, $MFe_3(SO_4)_2(OH)_6$, where M^+ represents either H_3O^+ (in hydronium jarosite), sodium (in natriojarosite) or potassium ions (in jarosite). Stoichiometrically, jarosites may be regarded as mixtures of MOH, iron(III) sulfate, iron(III) oxohydroxide and water, in the ratio 3:2:5:5. Such jarosites have been isolated from bacterial oxidation products. Jarosite formation is favoured at higher temperatures and is produced in pressure oxidations of iron minerals. Jarosites occur in supergene alteration zones (weathering by surface water) of pyrite deposits, in dolomitic zones in newly opened mine galleries, in sulfide-rich mine dumps, in hydrothermal springs and in sulfate soils. It is probable that such deposits are formed by bacterial action. Jarosite deposits are usually intergrown with limonite (which is a general term for goethite and other variously hydrated iron(III) oxides). As they weather they break down to give goethite, hematite or limonite. Although jarosite is reasonably stable at pH values less than three it is unstable with respect to goethite at higher pH values. Contact with water eventually causes the formation of goethite according to equation (7.1), which refers to potassium jarosite.

$$KFe_3(SO_4)_2(OH)_6 \rightarrow 3FeO(OH) + K^+ + 3H^+ + 2SO_4^{2-} \quad (7.1)$$

Equation (7.1) and the general observations of weathering of jarosite indicate that it is thermodynamically unstable with respect to the formation of goethite or hematite.

(ii) Iron(III) oxohydroxide, goethite, lepidocrocite and hematite.

As the pH value of an iron(III)-sulfate solution is increased, progressive hydrolysis of iron(III) takes place (Flynn, 1984) leading to the eventual predominance of the neutral $[Fe(H_2O)_3(OH)_3]$ complex. The details of the speciation of the hydrolysis products are dealt with in section 4.3.5.2 together with a discussion of the iron(III) speciation in the pH range from zero to two. At pH values above two the hydrolysis proceeds to an extent where precipitation of solids occurs. The manner of the formation of the eventually precipitated material is very complicated. Even if precipitation does not occur immediately there are complex changes in the system which are time dependent. The further hydrolysis of the mononuclear species (FeOH^{2+} and Fe(OH)$_2^+$) leads to the relatively rapid (one to two minutes at 25°C) formation of a red polymeric substance. This is the precursor of the orange/brown gelatinous solid which is sometimes referred to erroneously as iron(III) hydroxide. The initially produced red hydrolytic polymer consists of spherical colloidal particles with diameters of two to four nanometres. Such particles may be re-converted to the solution phase by the addition of acid, the process taking a time of between two to fifteen minutes. If the hydrolysis is allowed to continue the freshly produced spheres become hardened, with no change in size, to a form from which the reverse reaction is considerably slower (taking between three to thirty hours). The hardened spheres undergo an aging process, first joining up to give polymeric rods, then two-dimensional sheets and finally the three-dimensional solid material which forms the precipitate. In a solution in which there is no immediate precipitation the whole process can take as long as three to four years at ambient temperature, although it takes only a few days at 100°C. When the pH of an iron(III)-containing solution is raised to a value where precipitation is immediate the above processes would be affected by the lack of time to undergo the necessary ordering to give the thermodynamically most stable solid material. The gelatinous solid produced has been described as iron(III) hydroxide, but no compound with the Fe(OH)$_3$ stoichiometry seems to exist, and the better formulation as hydrated hematite, Fe$_2$O$_3$.xH$_2$O, is more appropriate for the precipitated material. When x = 1 the formula may be simplified to FeO(OH), iron(III) oxohydroxide, which is that of the mineral goethite. There is evidence that some of the FeO(OH) produced, which does not go through the prolonged polymerization described above, is the γ-form known as lepidocrocite.

The initially precipitated material inevitably possesses a high water content as is evident from its gelatinous nature. When separated from the supernatant solution and allowed to lose water by evaporation the eventual product is hematite, Fe$_2$O$_3$, although this depends upon the prevailing conditions. The change in standard Gibbs energy, $\Delta G^{\ominus}$, for the reaction:

$$2\text{FeO(OH)(s)} \rightarrow \text{Fe}_2\text{O}_3\text{(s)} + \text{H}_2\text{O(l)} \qquad (7.2)$$

has a small negative value (-7.4 kJ mol^{-1}) which implies that the formation of hematite would not take place in wet conditions. It is only formed from goethite by weathering in dry conditions where the product water can evaporate.

For the iron(III)-sulfate system it appears that the most stable compounds are goethite in wet conditions and hematite when dry. Jarosite is a possible intermediate under certain conditions. The interconversion rates are very low making it possible for jarosite, goethite and hematite to co-exist under various conditions for a considerable length of time. The solubility of iron(III) oxohydroxide is sufficient to produce a solution containing around 3 x 10^{-12} M iron(III) if the pH value of the solution is between six and eight.

7.4.1.2 Calcium hydroxide (or limestone) as precipitant

The treatment of sulfate-containing solutions with aqueous calcium ion produces calcium sulfate in the form of gypsum, $CaSO_4.2H_2O$. In addition to the considerations in the NaOH/KOH section above the only calcium-containing mineral which is relevant is calciocopiapite, $CaFe_4(SO_4)_6(OH)_2.19H_2O$. Stoichiometrically this can be viewed as a 1:2 mixture of calcium hydroxide and iron(III) sulfate. The theoretical presence of the potentially soluble iron(III) sulfate ingredients would make it unlikely that the compound would be produced. The precipitate from an iron(III)-sulfate effluent would be expected to be gypsum, together with the gelatinous iron(III) product as discussed in the previous section. Any use of lime for the neutralization of a solution inevitably leads to an excess of it in the solid product mixture.

7.4.2 Compounds containing iron(III) with arsenic(V) and/or sulfur(VI)

If arsenic(V) is present in the effluent solution, as is usually the case with bacterial oxidation, there are other compounds in addition to those of the previous section which have to be considered. Continuing with the above approach, known iron(III)-arsenate and iron(III)-sulfate-arsenate minerals are now considered.

7.4.2.1 Sodium (or potassium) hydroxide as precipitant

(i) Iron(III) arsenate minerals.

The two simple iron(III)-arsenate minerals which exist are scorodite, $FeAsO_4.2H_2O$, and kankite, $FeAsO_4.3.5H_2O$. Scorodite has been identified in the effluent treatment from pressure oxidation of arsenopyrite. There is no evidence for the production of kankite in precipitation reactions. The mineral is only found in one deposit in Kaňk, Czechoslovakia (Čech *et al.* 1976).

(ii) Basic iron(III) arsenate minerals.

There is a basic iron(III) arsenate, ferrisymplesite, $Fe_3(AsO_4)_2(OH)_3.5H_2O$, which is amorphous, and one containing potassium, pharmacosiderite, $KFe_4(AsO_4)_3(OH)_4.6\text{-}7H_2O$. Sodium pharmacosiderite has a variable sodium/potassium composition and is formulated as $(Na,K)_2Fe_4(AsO_4)_3(OH)_5.7H_2O$. Ferrisymplesite may be regarded as a 2:1 stoichiometric mixture of iron(III) arsenate and iron(III) oxohydroxide. The pharmacosiderites have stoichiometric compositions of 3:1:1 (iron(III) arsenate: iron(III) oxohydroxide: potassium hydroxide) and 3:1:2 (iron(III) arsenate: iron(III) oxohydroxide: sodium/potassium hydroxides).

(iii) Basic iron(III) arsenate sulfate minerals.

There are three basic iron(III)-arsenate sulfates; bukovskyite, $Fe_2AsO_4SO_4OH.7H_2O$, sarmienite, $Fe_2AsO_4SO_4OH.5H_2O$, and zykaite, $Fe_4(AsO_4)_3SO_4OH.15H_2O$. They may be regarded stoichiometrically as mixtures of iron(III) arsenate, iron(III) sulfate and iron(III) oxohydroxide with respective ratios of 3:1:1, 3:1:1 and 9:1:1. The presence of a stoichiometric proportion of what would be soluble iron(III) sulfate is a possible indication that they are not the compounds which represent ultimate thermodynamic stability. Although the thermodynamic data are not available it is likely that they would give iron(III) arsenate and iron(III) oxohydroxide by the reaction (e.g. for bukovskyite and sarmienite):

$$Fe_2AsO_4SO_4OH + 2OH^- \rightarrow FeAsO_4 + FeO(OH) + H_2O + SO_4^{2-} \qquad (7.3)$$

Bukovskyite and zykaite exist in an ancient dump in Czechoslovakia and are proposed to be products of supergene alterations of arsenopyrite and pyrite (Novák et al. 1967). There are no other known sources which is consistent with the conclusion that they are of only intermediate stability. Sarmienite was reportedly found in Argentina (Angelelli and Gordon, 1941).

As the pH of an iron(III)-sulfur(VI)-arsenic(V) aqueous system is increased it might be expected that basic iron(III) sulfate arsenates would be precipitated. These are thermodynamically less stable than the equivalent mixture of iron(III) arsenate and iron(III) oxohydroxide. The addition of a solution of NaOH or KOH at a pH of thirteen (0.1 M OH^-) to an acid effluent solution produces a local environment, around each drop of alkaline solution, with an excessively high pH value. After stirring has produced a more even mixture the solids produced may be far from being those corresponding to the equilibrium composition at the current pH of the system. In practice a gelatinous precipitate is produced which does contain iron(III) with sulfate and arsenate ions. The reorganization of such precipitates to form more stable products can be very slow. Solids of composition similar to the basic iron(III)-sulfate-arsenates which are found in nature would be expected to persist for long times before their conversion to thermodynamically more stable products. These are likely to be iron(III) arsenate and either iron(III) oxohydroxide (goethite) or iron(III) oxide (hematite), with any sulfate being eventually solubilized.

7.4.2.2 Calcium hydroxide (or limestone) as precipitant

(i) Gypsum

The sulfate content of the solution would be precipitated as gypsum as is the case with iron(III)-sulfate solutions discussed in section 7.2.1.2.

(ii) Calcium arsenates

Nishimura et al. (1987) have determined thermodynamic data for five calcium(II)-arsenic(V) compounds which are possible candidates for arsenic(V) stabilization. The compounds and their standard Gibbs energies of formation are given in Table 7.1. The compounds shown in Table 7.1 are in the order of their formation as the pH value of the system increases and are consistent with the predominant form of arsenic(V) acid (see section 4.3.3) in each pH region. The compound $Ca_5H_2(AsO_4)_4$ is possibly better formulated as containing the $HAsO_4^{2-}$ and AsO_4^{3-} ions: $Ca_5(HAsO_4)_2(AsO_4)_2$. In terms of the standard Gibbs energy of formation per mole of arsenic(V) the compound Ca_2AsO_4OH offers the greatest stability for arsenic(V). In a calcium(II)-arsenic(V)-sulfur(VI) system, it is precipitated in the pH range above ten. In practical situations it is unlikely that the pH value of the effluent solution would exceed ten. Providing that is the case then calcium arsenate, $Ca_3(AsO_4)_2$, is the most likely solid product. It is not, however, found in mineral form (nor is $CaAsO_4OH$), although there are two basic calcium arsenates which exist.

Table 7.1 The standard Gibbs energies of formation for some calcium(II)-arsenic(V) compounds

Compound	ΔG_f°/(kJ mol^{-1})	ΔG_f°/(kJ mol.As^{-1})
$Ca(H_2AsO_4)_2$	-2053.9	-1026.95
$CaHAsO_4$	-1287.4	-1287.4
$Ca_5H_2(AsO_4)_4$	-5636.7	-1409.2
$Ca_3(AsO_4)_2$	-3060.6	-1530.3
Ca_2AsO_4OH	-1987.8	-1987.8

(iii) Basic calcium-iron(III) arsenates

The two calcium-containing minerals which could be considered as model compounds for arsenic control are arseniosiderite, $Ca_3Fe_4(AsO_4)_4(OH)_6.3H_2O$, and kolfanite, $Ca_2Fe_3(AsO_4)_3(OH)_4$. Both of these compounds can be formulated as stoichiometric mixtures of calcium hydroxide and iron(III) arsenate, the ratios being 3:4 and 2:3 respectively. It is unlikely that such compounds would be thermodynamically very stable.

7.4.2.3 The relative stabilities of iron(III) and calcium(II) arsenates

Thermodynamic data are available for iron(III) and calcium(II) arsenates so that it is possible to postulate some reactions for their interconversion. In aqueous solution, at values of pH below 11, calcium(II) occurs as the hydrated calcium(II) ion, Ca^{2+}(aq). The equation representing the conversion of iron(III) arsenate to calcium arsenate by reaction with aqueous calcium(II) ion may be written as:

$$2FeAsO_4 + 3Ca^{2+} + 4H_2O \rightarrow Ca_3(AsO_4)_2 + 2FeO(OH) + 6H^+ \qquad (7.4)$$

which has a change in standard Gibbs energy of 168.6 kJ (mol.equation)$^{-1}$. This result implies that calcium arsenate is not relevant to the stabilization of arsenic(V) when iron(III) is present in the system. The position of equilibrium of reaction (7.4) is dependent upon the value of the pH of the system. In strongly alkaline conditions, where the calcium(II) ion is stable as the hydroxide (at pH values greater than 11), the reaction is best written as:

$$2FeAsO_4 + 3Ca(OH)_2 \rightarrow Ca_3(AsO_4)_2 + 2FeO(OH) + 2H_2O \qquad (7.5)$$

which has a change in standard Gibbs energy of -202 kJ mol^{-1}, indicating that the arsenic(V) is more stable in the form of calcium arsenate. The formation of calcium arsenate should be avoided because of its possible conversion to calcium carbonate by interaction with dissolved carbon dioxide. This would lead eventually to the complete solubilization of the arsenic(V) (Nishimura *et al.* 1987).

7.4.3 Possible compounds representing ultimate thermodynamic stability

Consideration of all the above compounds which are possibly precipitated from iron(III)-sulfate and iron(III)-sulfate-arsenate solutions by either sodium/potassium or calcium hydroxides indicates that (if their potentially soluble components are taken into account) there are only four compounds which may be taken as having considerable thermodynamic stability. These are iron(III) oxohydroxide (goethite), iron(III) oxide (hematite), iron(III) arsenate (scorodite), and calcium sulfate (gypsum). There are important considerations concerning the solubility and ultimate stability of iron(III) arsenate. Its solubility must comply with local environmental regulations about the permitted levels of iron and, in particular, arsenic.

7.5 THEORETICAL MODELLING OF PRECIPITATE STABILITIES

The speciation mathematics of section 4.3.5.3 includes the experimentally determined values for the formation constants of $FeAsO_4$(aq) and $Fe(OH)_3$(aq), these species being the supposed precursors of the precipitated iron(III) arsenate and iron(III) oxohydroxide respectively. The calculations also allow for the formation of the anions, $Fe(AsO_4)_2^{3-}$ and $Fe(OH)_4^-$ which are responsible for the solubility of iron(III) and arsenic(V) at higher values of pH. The speciation of soluble iron(III) complexes, at low values of pH, is shown in Fig. 4.5. Fig. 7.1 shows the variation with pH of the alpha values of (i) the soluble iron(III) complex ions (expressed as a total), and (ii) the four species mentioned above, for a solution which was originally 0.3 M in iron(III), 0.45 M in sulfur(VI) and 0.2 M in arsenic(V).

In reality the nature and composition of any solid initially precipitated may be different from the eventually (and thermodynamically more stable) produced substances. In this section it is assumed that the precipitates produced are the ones with ultimate thermodynamic stability (i.e. those corresponding to the equilibrium state).

As the pH increases, the first precipitate is iron(III) arsenate which has a maximum predominance at a pH value of 4.5. Any iron(III) remaining in solution after the formation of the iron(III) arsenate begins to be precipitated as hydroxide at pH values greater than 3.5. The formation and increasing predominance of the $Fe(AsO_4)_2^{3-}$ ion accounts for the solubilization of the iron(III) arsenate at pH values above five, becoming a maximum over the pH range 7-12. At pH values above 8.5 the iron(III) oxohydroxide becomes increasingly more solubilized with the formation of the $Fe(OH)_4^-$ ion. At pH values greater than twelve the $Fe(OH)_4^-$ ion predominates over the comparatively less stable $Fe(AsO_4)_2^{3-}$ ion.

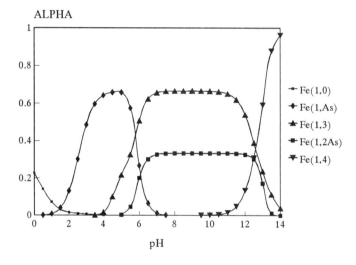

Fig. 7.1 Speciation of the major iron(III) species in a solution of 0.3 M iron(III), 0.45 M sulfur(VI) and 0.2 M arsenic(V)

Iron(III) arsenate is more soluble than iron(III) oxohydroxide to an extent which is of some environmental concern. In the pH range from two to six the observed solubility is due primarily to the aqueous $FeAsO_4$ ion pair complex which has a concentration of about 2×10^{-6} M (2 ppm). At pH values higher than five the formation of the bis-arsenatoiron(III) ion causes the solubility of arsenic(V) to increase as is implied by the data shown in Fig. 7.1.

The considerations so far, describe the solubilities of iron(III) and arsenic(V) in the absence of calcium ion, i.e. they are relevant to the use of sodium or potassium hydroxide as the neutralization agent. The use of lime as a precipitant does not alter the conclusions providing that a sufficient excess of it is avoided. Its use would cause the removal of the sulfur(VI) as gypsum. Excessive use would initiate the conversion of iron(III) arsenate to calcium arsenate (equation (7.5)) which would eventually become calcium carbonate, with the solubilization of the arsenic content, as a result of the interaction with atmospheric carbon dioxide.

The use of sodium (or potassium) hydroxide leads to the eventual formation of iron(III) arsenate and, providing the pH remains lower than six, allows any contact water to have an arsenic(V) concentration of no more than 2 ppm. It also avoids the solubilizing effect of carbon dioxide. Unless a more stable compound with a lower solubility is found it may well be that industry will have to argue a case for a possible long-term slow release of this level of arsenic(V). The neutralization of the effluent using sodium hydroxide, although a more expensive method than that using lime or limestone, would produce the relatively stable iron(III) arsenate, the remaining solution either being re-cycled (essentially consisting of dilute sulfuric acid) or being treated with lime to produce gypsum and water which could be allowed into the environment safely. This method would avoid the conversion of iron(III) arsenate to the undesirable calcium arsenate.

7.6 EXPERIMENTAL INVESTIGATIONS OF PRECIPITATED SOLIDS

Extensive work has been carried out upon the iron(III)/arsenic(V) system by Krause and Ettel (1985, 1989), Robins (1981, 1990), Khoe and Robins (1988) and Harris and Monette (1988). Large amounts of data have been produced by Nishimura et al. (1987) on the calcium(II)-sulfur(VI)-arsenic(V) and calcium(II)-arsenic(V) systems that are iron(III) free. They also present data which are concerned with the effect of atmospheric carbon dioxide upon the stability of the various calcium arsenates as a function of pH. Robins et al. regard the precipitates from iron(III)/arsenic(V) solutions as iron(III) oxohydroxide (or ferrihydrite), with the co-precipitated iron(III) arsenate on its surface. They deny the existence of basic iron(III) arsenates. It is certainly true that the formulae of the minerals bukovskyite and sarmienite can be formulated as mixtures of iron(III) arsenate, iron(III) sulfate and goethite (see section 7.2.2.1, (iii)) and, conversely, it is possible to represent iron(III) arsenate adsorbed on to a goethite surface as a layer of basic iron(III) arsenate. Whether this is merely a semantic argument or whether there are demonstrable structural and chemical differences between the two proposed types of precipitate will emerge as further work is carried out. Krause and Ettel report that the ideal [Fe(III)]/[As(V)] ratio in solution is 4.0 to give maximum solid stability, the solution containing only 0.3 mg L^{-1} (3 ppm) of arsenic. The use of sodium hydroxide gives a solid which is more stable (soluble arsenic = 0.008 mg L^{-1}) than that produced by the use of slaked lime. Robins et al. (1988) have shown that an iron/arsenic ratio of ten gives the observed maximum stability, although a value of only two was sufficient to reduce the solubility of arsenic to 0.2 mg L^{-1}. Hackl (1990) has shown that the precipitate from the neutralization of the effluent from the Salmita bacterial oxidation plant, when subjected to re-pulping in water at seventeen percent solids, produced a solution in which the initial concentration of arsenic was 6 mg L^{-1}. After a period of around one hundred and fifty days, in which the slurry had been continuously stirred (which included a re-pulping with water at day fifty five), the arsenic concentration became steady at 0.1 mg L^{-1}, the pH of the solution being 7.4. Hackl considered that the changes were due to the conversion of the supposed initial products, hydronium jarosite and hydrated iron(III) arsenate, to the thermodynamically more stable iron(III) oxohydroxide, gypsum and calcium arsenate. For this to occur there would have to be an excess of lime present.

Barrett *et al.* (1991a) have shown that the use of slaked lime slurry to neutralize bacterial oxidation effluent solution can lead to the inclusion of slaked lime in the solid product. This can have the effect of increasing the pH to greater than acceptable levels when it is eventually liberated (at pH values greater than six iron(III) and calcium arsenates become reasonably soluble). Using infra-red spectroscopy they have also shown that the sequence of precipitation, as the pH increases, is gypsum, calcium arsenate, and what they describe as basic iron(III) arsenate sulfates. The use of sodium hydroxide to neutralize the effluent produces an initial precipitate of a basic iron(III) arsenate sulfate followed by a basic iron(III) sulfate plus iron(III) oxohydroxide. This observation is consistent with those of Krause and Ettel's that the nature of the precipitating agent affects the stability of the precipitate. If sodium hydroxide were to be used as a precipitant it would have to be followed by a second stage in which the excess of sulfate would be removed by slaked lime treatment unless the dilute sulfuric acid solution was recycled. Barrett *et al.* (1991b) suggest that the advantage of the [Fe(III)]/[As(V)] ratio of greater than four is that any precipitate might be encapsulated by stable iron(III) hydroxide. Such a view is quite the opposite of that of Robins but since any iron(III)-arsenic(V) compounds are preferentially precipitated, before iron(III) oxohydroxide is produced, it should be considered.

7.7 CONCLUSIONS

The solution effluents from bacterial oxidation processes can be successfully treated with either slaked lime or the more expensive sodium hydroxide. Limestone could be used instead of lime, its lower reactivity possibly allowing a greater degree of control, leading to the initial production of thermodynamically more stable precipitates. Slaked lime treatment ultimately produces a mixture of gypsum, iron(III) arsenate and iron(III) oxohydroxide. The kinetics of arsenic(V) release from such mixtures need to be studied so that any release may be controlled and minimized.

If sodium hydroxide is used to neutralize the effluent solution a mixture of iron(III) arsenate and iron(III) oxohydroxide is produced.

Whichever neutralization agent is used there will be a low concentration of arsenic(V) produced in any contact water. With careful control its level is not likely to exceed 2 mg L^{-1} and Robins (1990), Robins et al. (1991) and Hopkin (1991) have suggested that complete containment should perhaps be regarded as an unattainable ideal and that environmental authorities should consider this in deciding what the permitted minimum level of arsenic(V) dispersal should be.

A very important and comprehensive review of the literature on the speciation of arsenic in the environment has been published by Cullen and Reimer (1989). They point out that arsenic is widespread in the environment and that its average level in sea water is 2 ppb but that in rivers can range from 0.1-75 ppb, with European rivers having a level of about 3.5 ppb. It is of note that the arsenic level in the Old Faithful hot spring in Yellowstone National Park is as much as 1275 ppb. This level is reasonably close to that expected (2000 ppb) for a solution in contact with pyrite and arsenopyrite which is undergoing bacterial oxidation. The sulfur spring activity in the Park is, of course, due to just that. Cullen and Reimer also discuss the role of bacteria in maintaining levels of arsenic(III) and arsenic(V) in the environment. Bacteria are known which can effect the oxidation of arsenic(III) and others which can effect the reduction of arsenic(V). There are species, ranging from various bacteria, fungi, yeasts, to fish and humans which have the facility to methylate arsenic(V) to give a variety of compounds including methylarsonic acid, $CH_3AsO(OH)_2$, and dimethylarsinic acid, $(CH_3)_2AsO(OH)$. Fish and shellfish generally contain fairly high levels of arsenic (e.g. 0.3 ppm in salmon to 14 ppm in plaice, 2 ppm in mussels to 50 ppm in octopus), the majority of their arsenic content being present as arsenobetaine, $(CH_3)_3As^+CH_2COO^-$. The normal daily human urinary excretion of arsenic is in the range between ten to fifty micrograms which is that expected from the arsenic content ingested from a normal diet. The figures for fish eaters are considerably higher with a daily excretion which could be seven to eight times greater than non-fish eaters. Fortunately for fish eaters the excess of ingested arsenic is present as arsenobetaine which passes into the urine without being metabolized and hence is no threat their health! Smelter workers who may be exposed to airborne arsenic trioxide eliminate the arsenic in the form of dimethylarsinic acid, the levels being around six times higher than those in control groups.

More work is required to determine the exact nature of the precipitated solids produced by the use of sodium, potassium or calcium hydroxides to neutralize the effluent solution, and to determine which compounds are initially precipitated and which are the ultimately formed and thermodynamically most stable compounds. The solubilities of all the precipitated materials and their eventual products with respect to arsenic release in the presence of atmospheric carbon dioxide should be recorded. Further research should be initiated with the view to producing less soluble compounds of arsenic(V) which would reduce the environmental impact, although with present practice this is acceptably low.

7.8 REFERENCES

Angelelli, V., & Gordon, S.G. (1941) *Acad. Sci. Phil. Not. Nat.*, **92**, 4.
Barrett, J., Hughes, M.N., Islam, A.N. & Simons, C. (1991a) *Randol Gold Forum, Cairns, 1991*, p.179.
Barrett, J., Hughes, M.N. & Islam, A.N. (1991b) *Biohydrometallurgy - 91, Portugal, 1991, in press.*
Čech, F., Jansa, J., & Novák, F. (1978) *N. Jahrbuch f. Mineralogie. Monatshefte* **H3**, 134.
Cullen, W.R., & Reimer, K.J. (1989) *Chem. Rev.*, **89**, 713.
Flynn, C.M. (1984) *ibid.*, **84**, 31.
Hackl, R.P. (1990) *Randol Gold Forum, Squaw Valley, 1990*, p.101.
Harris, G.B. & Monette, S. (1988) *Arsenic Metallurgy - Fundamentals and Applications Symposium, 1988 TMS Annual Meeting, Phoenix, Arizona, 1988*, Reddy, R.G., Hendrix, J.L & Queneau, P.B. (eds) TMS, 1987, p.469.
Hopkin, W. (1991) *Randol Gold Forum, Cairns, 1991*, p.191.
Khoe, G.H., and Robins, R.G. (1988) *J. Chem. Soc. Dalton Trans.*, p.2015.
Krause, E. & Ettel, V.A. (1985) *Impurity Control and Disposal Symposium - CIM Vancouver, 1985*, p.1.
Krause, E. & Ettel, V.A. (1989) *Hydromet.*, **22**, 311.
Nishimura, T., Itoh, C.T. & Tozawa, K. (1987) *Arsenic Metallurgy - Fundamentals and Applications Symposium, 1988 TMS Annual Meeting, Phoenix, Arizona, 1988*, Reddy, R.G., Hendrix, J.L. & Queneau, P.B. (eds) TMS, 1987, p.77.
Novák, F., Povondra, P. & Vtělenský, J. (1967) *Acta Univ. Carolinae, Geol.*, p.297.
Robins, R.G. (1981) *Metal. Trans.*, **12B**, 103.
Robins, R.G., Huang, J.C.Y., Nishimura, T. & Khoe, G.H. (1981) *ibid.*, p.99.
Robins, R.G. (1990) *Proceedings TMS/AIME Annual Meeting, Anaheim CA, 1990*, p.93.
Robins, R.G., Wong, P.L.M., Nishimura, T., Khoe, G.H. & Huang, J.C.Y. (1991) *Randol Gold Forum, Cairns, 1991*, p.197.

8

Economic factors

8.1 INTRODUCTION

The economics of any bacterial oxidation process are determined primarily by local factors. Every situation must be assessed individually and there are no general rules that can be applied to determine whether bacterial oxidation is the most appropriate technology. The costs for bacterial oxidation, as with alternative technologies, can be divided into the capital costs associated with construction and provision of services and the operating costs arising from the running of the process equipment and the supply of reagents and services. Generally the capital costs for bacterial oxidation costs are less than those of competing technologies and operating costs are comparable (Gilbert *et al.* 1988, Fraser *et al.* 1991).

8.2 CAPITAL COSTS

The capital costs for each of the different methods of application of bacterial oxidation processes normally increases in the order:

$$\text{dump} < \text{vat} \leq \text{heap} < \text{agitated reactor.}$$

Local factors have an influence for any specific application but the general trend still applies when considering the cost of establishing any bacterial oxidation process. Materials of construction and equipment are the major items affecting the capital costs and such costs are in the same relative proportion to the total costs in most mining areas.

The capital costs are highest for the agitated reactor bacterial oxidation system. The application of bacterial oxidation technology by the agitated reactor method is most comparable to alternative competing technologies (i.e. roasting and pressure oxidation) for difficult sulfidic ores and concentrates. Alternative technologies cannot be used in applications such as dumps, heaps, vats and *in situ* operations as can bacterial oxidation.

8.2.1 Major equipment

Major equipment required for agitated bacterial oxidation processes includes conventional tanks which use an impeller to suspend the solids and disperse the air, blowers or compressors to supply air, thickeners for solid/liquid separation and conventional slurry pumps to deliver and remove slurry from the plant. The pumps are similar to those used in flotation and carbon-in-pulp (CIP) leaching plants. Pressure oxidation processes require pressure vessels, flash vessels for the discharging slurry, expensive piping for oxygen delivery and high pressure pumps to inject the slurry. Roasting processes require stacks to disperse residual gases, a special pump to inject the high solids density slurry, circulating or fluidized beds to suspend the solids in the gas phase to allow the reactions to occur. Special blowers are necessary to introduce the gases into the sulfide bed at temperatures up to 700°C. By comparison, the equipment needed for bacterial oxidation is relatively simple and conventional.

8.2.2 Materials of construction

Bacterial oxidation requires acid resistant materials that are capable of withstanding relatively mild temperatures of 30 to 50°C. These materials of construction are of relatively low cost compared with the high temperature materials required in fluidized bed roasters and materials that have to withstand high temperatures, high acidity and oxygen pressures of pressure leaching technologies. The lower cost of materials can compensate partially for the additional quantity of materials resulting from the extended residence times that are typical of agitated bacterial oxidation processes.

Materials of construction that can be considered for a bacterial oxidation process include stainless steel, rubber-lined mild steel, polypropylene, high density polyethylene (HDPE), PVC, polyethylene and certain polyurethane coatings. All these acid resistant materials are used in conventional industrial processes and are capable of withstanding the operating temperatures used in a commercial bacterial oxidation process.

Pressure oxidation and roasting processes require specialized materials that can withstand severe oxidizing conditions and elevated temperatures. These materials are more difficult to obtain and need specialized techniques for their fabrication and forming.

8.2.3 Ancillary services

Other major costs in establishing a particular application are the services that are required to support a process. The capital costs of these services are not always included when comparisons of competing technologies are made but are an integral part of any project.

Roasting operations are no longer able to discharge sulphur dioxide-rich gases into the atmosphere and must recover the toxic component as sulfuric acid or gypsum. The cost of the acid recovery plant and the neutralization facilities are often of the same order of magnitude as the roaster. Previously, acid could be sold and viewed as a credit so that the costs of the ancillary services were not included in the capital cost of the roasting process. Sulfuric acid generated from roasting and smelting operations cannot always find a market and should not be assumed to be a credit against the cost of a project unless a ready and continuing market is available. Situations that involve arsenic trioxide are in the same position. The capital cost of arsenic removal from the discharge gases, usually by a bag house operation, can only be credited with the revenue from sale of high purity arsenic trioxide in the cases where markets exist for this product, there being only a small worldwide demand.

Pressure leaching capital costs are sometimes understated because the true cost of oxygen is disguised. In some parts of the world, it is practical to use a specialized oxygen producer to supply the tonnage of oxygen at a fixed cost. The final cost for oxygen will take into account the cost of the specialized producer constructing a plant in a specific area but may have the benefit of a larger scale than might be provided by the pressure oxidation process alone. Many mining areas are remote and do not have this benefit of an increased scale of operation for oxygen generation and should have the full cost of an oxygen plant included in the capital costs of the process.

Bacterial oxidation requires limited support services. Aeration is a major factor in the capital cost but the air can be provided in many cases by low cost blowers rather than higher cost air compressors. Temperature control can be optimized by balancing the size of the tanks with the amount of energy that must be removed or added to the process. Power for agitation which is translated into electrical installation and distribution can be a major cost but is no higher than for alternative technologies. It may be optimized by selection of appropriated agitators which balance reaction rates with power usage.

The civil engineering input associated with a bacterial oxidation plant can be more extensive than that required for roasting and pressure leaching, because of the larger areas required for the volume of tanks needed to provide the residence time for the process. The contribution of this civil engineering component to the overall capital cost will depend on the terrain in which the plant is constructed and the costs associated with earth moving and concrete foundations. The additional cost associated with installing specialized equipment and materials for the alternative technologies may equate with the additional civil engineering cost for bacterial oxidation.

8.3 OPERATING COSTS

The operating costs for a bacterial oxidation process include the major categories of power, reagents, services and labour. There are no saleable products from the process except the metals extracted. The acid generated is low grade and contaminated with dissolved metals and salts. The arsenic and other by-products are also contained in the waste streams and are not normally economically recoverable.

Power costs for the competing technologies are reported to be comparable (Gilbert et al. 1988, Fraser et al. 1991). The longer residence times of bacterial oxidation negate the benefit of a lower power intensity in the process. It is not clear whether the full costs of services have been included in the reported comparisons. The power costs for bacterial oxidation processes include agitation of the slurry, injection and dispersion of the air needed for oxidation, and power for services such as pumping, instrumentation and control. Power for roasting operations includes air to support the roasting operation, dewatering and pumping costs, power for the acid production and arsenic recovery plants, and instrumentation and control. The power costs for pressure oxidation processes should include power for agitation and gas dispersion, for pumping, for oxygen generation and storage, for temperature control, and for instrumentation and control.

Reagents represent a major proportion of the operating cost and are dominated by acid for control of initial acidity and lime or limestone for neutralization of acid waste streams before discharge. Acid generated from the oxidation of sulfide minerals must be neutralized and dissolved metals precipitated before solution can be discharged. The sulphur dioxide produced in roasting can be converted to sulfuric acid. However, the quantity of acid to be neutralized remains the same for any level of sulfide oxidation by any of the alternative processes. In the cases where significant arsenic is present in the sulfide minerals, roasting has additional costs for an oxidant for arsenic and for iron sulfate to form a stable precipitate. Arsenic in waste streams from bacterial oxidation and pressure oxidation is normally stable when precipitated with other dissolved metals, mostly iron, contained in the discharge solution. The chemistry of arsenic disposal is discussed in Chapter 7.

Roasting technology has had the benefit of being able to reduce its operating costs by credits from the sale of by-product acid and arsenic trioxide where true commercial markets existed. More recently, roasting operations have had to contend with neutralizing acid wastes and stabilizing arsenic products prior to disposal in containment dams. The neutralization costs are comparable to bacterial oxidation and pressure leaching because the quantities of acid are similar in most cases. Cost associated with arsenic stabilization are considerably higher for roasting operations because of the additional chemical reagents needed to convert arsenic trioxide to a stable iron arsenate residue.

Air requirements for bacterial oxidation are balanced by the air that must be injected into roasting to promote oxidation and the pressurized oxygen injection necessary for pressure oxidation. In whole ore roasting where arsenic is present in significant quantities, oxygen must be provided and circulated in the bed to stabilized the arsenic in the residue. Low pressure air from blowers, all that is required for bacterial oxidation, is considerable less expensive than either air to form a fluidized bed or pure oxygen produced from air separation processes.

Labour costs to operate and maintain a process and its services is determined partly by the geographical location of the plant. Labour is relatively expensive in developed and industrialized countries where there is a demand for labour to operate industrial plants. In contrast, there is an excess of labour in many developing countries and the cost of labour is relatively inexpensive. Labour intensive processes would benefit from being located in a developing country where abundant labour is available at low cost. The labour costs are also influenced by the need to operate, control and maintain the process equipment. The more conventional and less complex the process, the lower the level of skills required by the labour force. Bacterial oxidation technology is not a labour intensive process and has modest plant operator and maintenance skill requirements.

Bacterial oxidation is a relatively simple technology which does not require significant instrumentation or sampling to provide high metal recovery. The services that must be provided for assaying, instrument engineering and maintenance are usually less than for the alternative processes treating similar sulfide materials. The lower levels of these services required by the bacterial oxidation process reduce the operating costs.

A minimum of process instrumentation is used, only pH, dissolved oxygen and temperature measurement. These instruments can be connected to controllers to automatically adjust the process conditions or, because the process is relatively slow compared with alternative technology, the adjustments can be made manually by the operators. A fluctuation away from the design operating conditions lasting several hours represents only a small proportion of the total treatment time which is measured in days.

Bacterial oxidation processes use mostly conventional equipment such as agitated tanks, thickeners for solid/liquid separation, and slurry pumps. The equipment is similar to that used in most mineral processing operations world-wide and is not labour intensive to operate. Plant operators are responsible for maintaining the supply of reagents, controlling the pumping of slurry into the process, and keeping the process parameters within the design range by adjustment of valves controlling air flow and water flow for temperature control.

The conventional design of the process equipment for bacterial oxidation and the fact that the process operates at temperatures and pressures close to ambient means that the maintenance costs of the plant are low compared with alternative technology such as roasting and pressure oxidation. The materials used in the components of the plant do not have to withstand the high temperatures and extremely corrosive environments that exist in these other processes. Maintenance personnel do not have to cool or depressurize equipment before repairs so the time required for maintenance and services should be correspondingly shorter.

8.4 BACTERIAL OXIDATION PROCESS COSTS

A number of cost comparisons between bacterial oxidation technology and alternative technology have been published. Some of the comparisons refer to specific, well defined cases so that the capital and operating costs are meaningful. Other comparisons refer to general, hypothetical examples so that the capital and operating cost data cannot be related to any real situation. All of the published cost comparisons are for treatment processes for refractory gold concentrates.

Some of the comparisons of costs for alterative processing technology for sulfide minerals give only relative numbers and make assumptions about the treatment process, the gold recovery and the sale of by-products. Kontopoulos and Stefanakis (1988) performed one of the more useful studies and published the data in Table 8.1. The data refer to a process treating an arsenopyrite/pyrite concentrate at the rate of 100,000 tonnes per annum.

Table 8.1 Relative capital and operating costs and revenues for a 100,000 tpa plant

	Roasting	Pressure oxidation	Bacterial oxidation
Capital cost/$M	2.8	2.5	1.0
Operating costs* $M per annum	7.0	11.4	9.0
Revenues $M per annum	32.0	36.7	36.0

*Excluding the cost of the concentrate

The comparative capital costs for the alternative processes have included estimated residence times, assumed arsenic handling and disposal costs, and used projected reactor costs. The operating costs included assumptions on process conditions, waste disposal and sale of by-products. Revenues were based on estimated recoveries for precious metals, partly based on laboratory and small scale pilot testing.

Bruynsteyn et al. (1986) published capital and operating costs for a 100 tonnes per day plant in northern Ontario, Canada treating a pyrite/arsenopyrite concentrate. Roasting, pressure oxidation and a proprietary bacterial oxidation process were compared in an engineering study based on pilot plant test data. The capital costs were divided into direct costs, EPCM (Engineering, Procurement, Construction and Management) and contingency. The operating costs for bacterial oxidation were reported for five major categories; reagents, labour, spares, consumables and power. Only total operating costs were reported for the alternative process. Table 8.2 summarizes the comparative capital and operating cost data. These data give a more useful comparison of capital costs because the actual dollar values are given. Other operators interested in the technology can adapt these total costs to situations elsewhere to obtain an order of magnitude cost for similar processes at other locations.

Table 8.2 Cost comparison (US$) of alternative processes

Method of mineral treatment	Capital cost/$M	Operating cost /($ tonne^{-1})
Roasting	4.89	55.90
Pressure oxidation	7.48	42.79
BIOTANKLEACH	3.78	43.09

Gilbert et al. (1988) published a comprehensive comparison of capital and operating costs for roasting and bacterial oxidation of pyrite based on results from an experimental program. The comparison of costs for each technology included revenue generated from each process and an economic analysis. The example used was a 150 tonne per day plant operating over a 10 year life in a North American situation. The pyrite concentrate contained 13.4 g gold/t, 91.1 g silver/t and 1% copper.

The capital cost included the categories of equipment, installation, piping, insulation instrumentation, electrical, building, engineering, construction, contractors fees, contingency and site related expenses. This breakdown of capital costs facilitates the translation of the comparison to other situations because allowance can be made for the influence of local factors on each of the cost categories rather than just on a combined total.

The operating costs were treated in the same level of detail. The categories were reagents, energy, labour, maintenance and contingency. These cost categories could also be adjusted to another situation to provide a reasonable initial comparison without the major expense of a feasibility study on all technology options.

The capital costs for roasting and bacterial oxidation were $17.48 million and $8.37 million (1986) respectively. The annual operating costs were $3.52 million and $3.85 million respectively. Annual revenues, including gold, silver, copper and steam credits, were $5.74 million for roasting and $3.85 million for bacterial oxidation. The economic analysis of the costs and the revenue is summarized in Table 8.3. The economics for bacterial oxidation for the situation examined by Gilbert et al. were more favourable than for conventional fluidized bed concentrate roasting. The economics of bacterial oxidation were reported to improve further over roasting with an increase in either gold price or in gold grade in the concentrate.

A series of studies examining roasting, pressure oxidation and bacterial oxidation at different sulphur levels and plant capacities has been published by Litz and Carter (1988), Carter and Litz (1990), and Litz et al. (1990). The example chosen is based in North America using a pyrite concentrate. The cost calculations were not based on specific test work but used assumed process parameters based on other published information. These studies demonstrated the changes in the capital and operating costs as the scale of the operation varied. Comparisons of costs for each technology with and without a flotation concentration step were also included. Only one of these comparisons recognized the important advantage that bacterial oxidation can have over the alternative processes. This advantage is that complete oxidation of the sulfide mineral in not always essential for high gold extractions (see Fig. 2.2). None of the other processes for treatment of refractory gold has reported the ability to oxidize the sulfide mineral partially and obtain high gold extractions.

Table 8.3 Comparative Economic Summary ($000) (mid-1986 U.S. Dollars)

	Roasting	Bacterial oxidation
Total revenue	5741	5663
Operating costs	3520	385
Capital costs	17483	8374
Operating profit	2221	1810
Net present value @ 12%	(6163)	(676)
Simple payback	8.7 years	6.2 years
Internal Rate of Return	2.6%	10.0%

These theoretical studies were useful in showing that the alternative processes were sensitive to variations in scale and that this variation should be considered in any feasibility study based on experimental data for an actual application. More importantly, these comparisons have examined the potential for partial sulfide oxidation using bacterial technology and showed that there were significant cost advantages where complete oxidation was not required. The extent of sulfide reaction necessary for high metals recovery is specific to each potential application and can only be established by test work on representative samples. The optimum level of oxidation may vary with depth and source of ore, even though the sulfide material may originate from the same mine. Comprehensive testing is essential to insure that the bacterial oxidation process is designed for the range of sulfide mineral to be produced at a given operation.

Each of the published studies on bacterial oxidation technology costs serves to reinforce the points already made. Costs are determined by factors specific to each potential application and it is not possible to translate costs from another situation and obtain an accurate assessment of capital and operating costs. Most published information does not contain sufficient detail to allow local factors to be applied to give other than order of magnitude estimates for a proposed application.

Detailed testing of representative samples for a specific application of bacterial oxidation is needed to obtain design ranges for all of the operating parameters. An engineering feasibility study specific to that application can be completed once the design criteria have been established. Site based pilot plant testing should be performed if greater confidence is required in the performance of any technology at a specific site and bacterial oxidation is not an exception.

8.4 REFERENCES

Bruynesteyn, A., Hackl, R.P. & Wright, F. (1986) *Gold 100, Proceedings of the International Conference on Gold, Extractive Metallurgy of Gold, Johannesburg, SAIMM, 1986*, Volume 2, p 353.

Carter, R.W. & Litz, J.E. (1990) *unpublished work presented at the EPD Congress 90, TMS Annual Meeting, Anaheim, USA, 1990.*

Fraser, K.S., R.H. Walton, R.H. & Wells, J.A. (1991) *Minerals Engng* **4** (7-11), 1029.

Gilbert, S.R., Bounds, C.O. & Ice, R.R. (1988) CIM Bulletin, **81**, 89.

Kontopoulis, A. & Stefanakis, M. (1988) *Perth International Gold Conference, Randol International Ltd, 1988*, p 157.

Litz, J.E. & Carter, R.W. (1988) *ibid.* p 133.

Litz, J.E., Carter, R.W. & Kenney, C.W. (1990) *Randol Gold Forum, Squaw Valley, Randol International, 1990*, p 39.

9

Analytical methods

9.1 INTRODUCTION

During the laboratory study and plant operation of the bacterial oxidation of metal sulfides it may be necessary to use one or more of a range of analytical methods. As part of the preliminary assessment of the suitability of material for bacterial oxidation, ores and concentrates have to be analyzed for metal content, including that of potentially toxic metals, by methods that are routinely used in the mining industry and in other laboratories. It may also be necessary to determine the concentrations of metals or metalloids (e.g. arsenic) during the bacterial oxidation process itself, firstly to assess the extent of solubilization of the mineral (and indirectly the viability of the culture) and secondly to give early warning of the build up of potentially toxic species (if these are present in the concentrate). In addition, analysis of the metal and/or metalloid content of waste liquors from cultures is necessary to ensure that the appropriate disposal methods are used. A number of criteria will determine the choice of analytical techniques in these operations, but it is important to distinguish between those that analyze for the total element, irrespective of the chemical forms in which it may be present, and those that allow the determination of particular chemical species. For example, the use of atomic absorption spectroscopy allows the analysis for total arsenic in a supernatant solution from a culture oxidizing arsenopyrite, but does not distinguish between arsenic(V) and the more toxic arsenic(III). The latter species can be determined accurately by polarographic methods.

As implied above, it may be desirable to monitor the viability of the culture, particularly when studying the bacterial oxidation of a new material. The extent of growth of the culture is often related to the amount of mineral solubilized, that is proportional to the concentration of iron or other metallic species present in the supernatant solution. The rate of solubilization of the mineral then approximates to the rate of growth of the culture. The decrease in pH value of the growth medium similarly provides an estimate of the extent of growth of a culture during the bacterial oxidation of sulfides. These methods are commonly used to monitor the state of cultures but are obviously imprecise and crude. It would, for example, be more informative to measure concentrations of iron(II) and iron(III) rather than total iron, while the extent of growth of the culture would better be determined by direct measurement of the amount of biomass in the reactor. It should also be noted that the relative distribution of the biomass between the mineral surface and the medium is probably a factor of particular importance for ensuring rapid and efficient bacterial oxidation of the mineral. Growth of bacterial cultures is traditionally monitored by measurement of optical density at a wavelength of around 600 nm (i.e. by light-scattering) or by the use of particle size analyzers. These methods fail in the presence of mineral particles and so alternative approaches are needed.

In this chapter, the methods available for the determination of total metal concentration and particular metallic species will be summarized while methods for following the growth of the culture will be discussed more fully. It should be noted that during analysis for metals it is not usually necessary to approach the detection limits for the method in use and so detection limit may not be an important factor in the selection of a technique. The textbook by Skoog *et al.* (1992) is recommended as a general reference for analytical procedures.

9.2 TOTAL CONCENTRATIONS OF ELEMENTS

Several techniques are commonly used, each having advantages and disadvantages, for the determination of the total concentrations of elements present in a sample. Probably the most commonly used methods are atomic absorption spectroscopy (AAS) and inductively coupled plasma atomic emission spectroscopy (ICPAES). The plasma source has been combined with mass spectroscopic detection (ICPSMS) to give significantly improved detection limits. Techniques allowing multi-element analysis are becoming more important and offer obvious benefits. An advantage of ICPSMS is that remarkably similar detection limits are obtained for the majority of elements, thus eliminating the need for repeated dilutions to determine all elements present in a sample. In contrast, detection limits for ICPAES are highly element-dependent.

9.2.1 Atomic absorption spectroscopy: flame and electrothermal methods

Detailed methods for the determination of many elements by atomic absorption spectroscopy are available in instrument manufacturers' handbooks. The technique utilizes a light source (visible or ultra-violet) which emits the atomic spectrum of the element to be analyzed. The sample is atomized in the light path giving atoms in the ground state which then absorb the incident radiation, causing a decrease in the intensity of the light beam reaching the detector. Use of a series of standard solutions to calibrate the instrument then allows the measurement of the concentration of the unknown solution. The calibration is only linear over a limited concentration range and all measurements must be carried out within this range. Alternatively, the method of standard additions may be used.

The two most common methods for sample atomization are the flame and the graphite furnace. In the former technique, the analyte solution is sprayed into the flame via a nebulizer. An air-acetylene flame is usually hot enough (1300°C) to atomize the sample but a nitrous oxide-acetylene flame may be necessary for refractory materials. The use of electrically heated graphite furnaces for atomization gives detection limits two orders of magnitude lower than those obtained using flame atomization. The sample is placed in the graphite furnace which lies on the light path. This operation is usually the major source of imprecision in this technique. The temperature of the graphite furnace is then raised in stages to allow the evaporation of the solvent, the ashing of the sample and finally the atomization of the required element. The temperature may rise to 3000°C in some determinations. The optimization of the heating programme represents an essential feature of the development of the analytical procedure: thus the heating rate during atomization must be sufficient to maintain a high density of atoms in the light path. An important advantage of the graphite furnace is the small sample size required (usually 10-50 µL), in contrast to the few millilitres required in the flame method. The use of larger samples in the graphite furnace allows lower detection limits to be achieved.

The main problem in atomic absorption spectroscopy is that of interference from chemical and biological material present in the sample, in particular from molecular absorption in the ultra-violet range which leads to spuriously high concentrations of metal in the sample. Several alternative methods for providing background correction to deal with this problem are now available. Other errors, both positive and negative, may arise during atomic absorption analysis but these are usually described in the methods given in instrument manufacturers' handbooks. It is most important that the validity of the analytical method be checked carefully before it is applied extensively, ideally by using an independent technique to confirm the result.

Atomic absorption analysis for metals can usually be carried out satisfactorily on aqueous samples or on acid digests of cells.

9.2.2 Atomic emission spectroscopy

This technique differs from atomic absorption spectroscopy in that it is dependent upon the constituent atoms of the sample which are electronically excited. The radiation emitted, as the result of the return of the excited species to the ground state, is measured and is linearly related to the concentration of the element over a wide range. Atomization and electronic excitation of the sample may be accomplished by conventional flame techniques, but recently atomic emission spectrometers have utilized inductively coupled plasma (ICP) sources for atomization. The high temperatures used (between five and ten thousand degrees Celsius) has meant that ICPAES is remarkably free from the problem of chemical interference. ICPAES is especially useful for multi-element analysis. About seventy five elements may be determined with precision and detection limits superior to flame atomic absorption spectroscopy.

9.2.3 Inductively coupled plasma mass spectrometry

The use of inductively coupled plasma has proved to be an excellent source of inorganic ions for determination by mass spectrometry. This method provides high sensitivity, with detection limits in the range 0.02-0.7 ppb (µg L^{-1}), the majority of elements having very similar values.

9.3 MEASUREMENTS OF SPECIFIC CHEMICAL SPECIES

9.3.1 pH measurement

The measurement of pH is an extremely important and well-studied technique. Bates (1955) has published a comprehensive account of the subject, a more recent discussion being given by Kristensen *et al.* (1991). The pH value of a solution usually consists of the use of a glass electrode and a saturated calomel reference electrode, often combined as a single unit. The effectiveness of the glass electrode depends upon the observation that, when immersed in a solution, the potential developed is a linear function of the pH as expressed by the Nernst-type equation:

$$E_{glass} = -0.0592.\mathrm{pH} + c \qquad (9.1)$$

where c is a constant characteristic of the particular electrode. The potential developed by a cell consisting of a glass/calomel electrode system in contact with a solution is given, in practice, by the equation:

$$E_{cell} = E_{glass} - E_{calomel} + E_j \qquad (9.2)$$

where E_j represents a summation of the junction potentials and which is not measurable. A combination of equations (9.1) and (9.2) gives:

$$E_{cell} = -0.0592.\mathrm{pH} - E_{calomel} + E_j - c \qquad (9.3)$$

which, by collecting the last three terms together and calling the sum C, may be simplified to give:

$$E_{cell} = -0.0592.\mathrm{pH} + C \qquad (9.4)$$

If the pH meter/cell combination is calibrated by using a solution of accepted pH value, pH_{cal}, equation (9.4) may be written as:

$$E_{cal} = -0.0592.\mathrm{pH}_{cal} + C \qquad (9.5)$$

If the electrode combination is then placed into a sample solution of unknown pH (pH_{sample}) the cell potential is given by the equation:

$$E_{sample} = -0.0592 \cdot pH_{sample} + C \qquad (9.6)$$

The collection of unknown constants, C, for any given system, may be eliminated by subtracting equation (9.5) from equation (9.6) to give:

$$E_{sample} - E_{cal} = -0.0592(pH_{sample} - pH_{cal}) \qquad (9.7)$$

The equation may be rearranged to give the value of the pH of the sample as:

$$pH_{sample} = pH_{cal} + (E_{cal} - E_{sample})/0.0592 \qquad (9.8)$$

A properly calibrated pH meter/electrode system makes use of equation (9.8) and allows the pH readings to be used with confidence, the actual errors in reading the pH value being in the region of ±0.03 units.

Glass electrodes commonly behave satisfactorily over the pH range from one to ten. At higher pH values the electrode is subject to an alkaline error which causes readings to be lower than they should be. At pH values lower than one there is an acid error which results in incorrectly high readings.

There are some important practical points to be borne in mind when taking pH measurements.

(i) Glass electrodes should be soaked in distilled water, or a buffer solution, for some hours before use. If allowed to go dry they must be soaked for at least twelve hours. It is good practice never to allow electrodes to dry out, and to leave the pH meter/electrode system switched on constantly.

(ii) The potential of the glass electrode for a particular pH value is dependent upon the temperature of the solution (the Nernst 'slope' being given by $-RT/F$, which has a value of - 0.0592 at 298 K). Most meters incorporate an automatic temperature compensator.

(iii) The pH meter/electrode system must be calibrated by using buffer solutions with accepted pH values. The buffer solutions which are recommended for the pH range one to five are hydrochloric acid/sodium citrate mixtures. Solutions containing various ratios of hydrochloric acid and potassium chloride are recommended (Bower and Bates, 1955) for pH values in the range between zero and two.

(iv) In solutions of low pH and especially when precipitates are part of the system the pH meter/electrode system should be allowed to equilibrate with each sample for at least fifteen minutes before a reliable reading can be taken.

9.3.2 Polarographic methods

The most commonly used polarographic technique is that of differential pulse polarography, which allows the simultaneous determination of a number of different metals in solution, and which may allow differentiation between species containing the same metal ion. It is of particular application in the determination of arsenic(III) in the presence of arsenic(V). Greater sensitivity is obtained using stripping voltammetry. In this technique the metal ion(s) is reduced at the hanging drop mercury electrode, usually forming an amalgam. This is the concentration stage and may be continued until an appropriate fraction of the total analyte is reduced. The analytical stage involves the stripping of the metal from the amalgam by oxidation, the magnitude of the peak current being proportional to the original concentration of the species in solution.

9.3.3 Ultra-violet and visible techniques, including colorimetric methods

Colorimetric methods are useful for measuring the concentrations of iron(III) or iron(II) species in each other's presence, requiring the use of relatively simple equipment. Iron(III) may be determined by the absorption of the thiocyanatoiron(III) complex, $[FeSCN]^{2+}$. The absorption maximum occurs at a wavelength of 460 nm, the molar absorptivity (or molar absorption coefficient), ε, having a value of 4800 L mol^{-1} cm^{-1}. Alternatively, iron(III) may be analyzed directly in sulfuric acid solutions by making use of the charge-transfer band with an absorption maximum at 305 nm and ε of 2250 L mol^{-1} cm^{-1} (Barrett, 1959). Iron(II) may be determined by the absorption of the tris-1,10-phenanthrolineiron(II) complex, $[Fe(phen)_3]^{2+}$, (phen = 1,10-phenanthroline) at 515 nm where the ε value is 11500 L mol^{-1} cm^{-1}.

9.3.4 Ion chromatography

This is an important relatively new technique for the separation and quantitative determination of charged species. In this technique the analyte, which may contain several species requiring analysis (cations, for example), is loaded on to a column containing a suitable ion exchange resin (a cation exchange resin in this case). The cations are then eluted from the column by pumping through a solution containing a cation that competes with the analyte cations for the negatively charged groups on the resin. The cations are released from the resin and as the various cations have different affinities for the resin separation occurs during the elution process. The appearance of ions in the eluant from the column can be identified by various techniques, often by conductivity (although precautions have to be taken to allow for the conductivity of the eluting electrolyte), and their concentration determined. Alternative detector systems may involve ultraviolet/visible, fluorescence or atomic absorption spectroscopy. Ion chromatography allows the rapid determination of inorganic charged species, often at concentrations less than 50 ppb for sample volumes less than 1 mL. These sensitivities sometimes compare well with those found for flame atomic absorption analysis, but may be improved by the use of larger samples or by a preconcentration step.

9.3.5 Gold analysis

It is essential, in monitoring the effectiveness of bacterial oxidation to enhance the recovery of gold from refractory concentrates, to use an accurate method for gold analysis. Fire assay is a traditionally acceptable and accurate method. An alternative method is to use atomic absorption spectroscopy. The gold content of the concentrate sample is dissolved in *aqua regia* (a 1:3 by volume mixture of concentrated nitric and hydrochloric acids). This treatment causes the gold content to be oxidized to the gold(III) complex, $[AuCl_4]^-$, which is extractable into a water immiscible organic solvent such as 2-methyl-4-pentanone (methyl isobutyl ketone, MIBK) or 2,6-dimethyl-4-heptanone (di-isobutyl ketone, DIBK). The *aqua regia* treatment causes the dissolution of all the components of a mineral concentrate except for any silicious gangue material. It is essential to wash the organic layer with 1 M hydrochloric acid to remove any iron(III), which is extracted as the $[FeCl_4]^-$ complex, as this interferes with the atomic absorption spectrometric analysis of the gold.

9.4 MEASUREMENTS OF BIOMASS

9.4.1 General comments: measurement of biomass on and off the mineral

As described in Chapter 3, chemolithoautotrophic bacteria may either be grown in a medium containing Fe(II) and/or a reduced sulfur compound or in a medium containing pyrite or another sulfide ore. In the former case the growth of the organism may be followed directly by measuring the optical density at 600 nm. However this method is not possible in cases where mineral substrate is also present. Microscopic counting has been used to evaluate bacterial numbers in the supernatant solution from such systems but this method does not allow cells attached to the mineral surfaces to be counted and does not distinguish between active and inactive cells. The uptake of $^{14}CO_2$ and ^{32}P-labelled compounds from the medium has also been used as a measurement of bacterial activity but this approach is probably only realistic on a laboratory scale.

Estimation of biomass has been attempted by analyzing samples of the culture for elements or molecules that must unambiguously be of microbiological origin, such as total organic carbon, organic nitrogen, protein, DNA or ATP. Protein, DNA and ATP can all be determined by standard, well-tried methods, but the use of these methods in monitoring growth of chemolithoautotrophs is complicated by the low yield of cells compared to those generally found for other bacteria and by interference from the solubilized mineral. Limited success has been achieved using protein assays but not for DNA or ATP at the present time. The current assays for DNA suffer from lack of sensitivity.

Estimates of the relative amounts of biomass suspended in the culture supernatant solution and adsorbed on the surface of minerals may be obtained by analyzing for protein in a sample of the culture supernatant from which the mineral has been separated and comparing this value with the result obtained from a sample which has been treated with detergent (sodium dodecyl sulfate) or some other reagent or sonicated to remove cells from the mineral surfaces (Section 9.4.3).

9.4.2 Analysis of protein

Protein concentration is usually determined by the well-known Lowry method (Lowry *et al.* 1957), which involves the formation of a heteropolymolybdenum blue, or otherwise by the use of other reagents such as Coomassie blue or bicinchinonic acid. The Lowry method is subject to interference from many species including Fe(II) and Fe(III). In reported methods, samples of the culture are therefore treated with alkali to precipitate iron(II) and iron(III) before proceeding with the normal Lowry assay on the supernatant solution. However, heating bacterially oxidized pyrite with alkali at temperatures greater than 50°C (for example, to remove cells from the mineral surface) leads to the formation of an intense yellow solution containing polysulfides which also causes interference. This problem does not appear to be recognized fully and there is no doubt that some methods reported for protein assay of chemolithoautotrophs give spurious results. This problem may be partly alleviated by oxidizing any sulfide to sulfate ion with potassium iodate (Ewart, 1990) but the iodide formed also interferes with the Lowry method. A further complication may arise during the bacterial oxidation of arsenopyrite due to the presence of arsenic(III) which can also serve as a reducing agent for the formation of the heteropolymolybdenum blue (Ewart, 1990). Indeed the Lowry method for protein can be used for the quantitative determination of the arsenic(III) concentration, the method being additionally sensitive to a range of reducing agents.

9.4.3 Differentiation between free cells and mineral-bound cells and between growing and non-growing bacteria

As noted earlier it may be extremely important to measure biomass on and off mineral surfaces. The measurement of total biomass requires the removal of cells from the mineral before the protein assay described above is carried out. Subtraction from this value of the biomass determined for the supernatant solution will then give the surface-bound biomass. Removal of the bacteria from the mineral surface may be accomplished by treatment with detergent or by boiling in alkali (Chang and Myerson, 1982) or by sonication (Ewart, 1990). Ewart has reported protocols for measurement of total biomass (protein) and free biomass in the presence of either pyrite or arsenopyrite. Shrestha (1988) has used the Coomassie blue protein assay for protein to estimate total and supernatant biomass.

Epifluorescence microscopy, using certain nucleic acid dyes that fluoresce under blue light, has also been used to distinguish between *T. ferrooxidans* attached to particles and free cells, and has been applied to the determination of the relative amounts of free and bound cells during the oxidation of pyrite (Yeh *et al.* 1987).

Populations of pyrite-oxidizing microorganisms have been estimated by the use of fluorescent antibodies (Apel *et al.* 1976, Baker and Mills, 1982). The method of Apel *et al.* for determining *Thiobacillus ferrooxidans* has been extended (Muyzer *et al.* 1987) to include a DNA-fluorescence staining technique which has allowed the selective enumeration of *Thiobacillus ferrooxidans* in a population of acidophilic bacteria. Unfortunately these methods are not very sensitive, requiring a culture density of 10^8 cells per gram of culture. The use of fluorescent antibodies has also been extended (Baker and Mills, 1982) to distinguish between growing and non-growing bacteria.

9.5 REFERENCES

Apel, W.A., Dugan, P.R., Filppi, J.A. & Rheins, M.S. (1976) *Appl. Environ. Microbiol.*, **32**, 159.
Baker, K.H. & Mills, A.L. (1982) *Appl. Environ. Microbiol.*, **43**, 338.
Bates, R.W. (1955) *Determination of pH*. 2nd edn John Wiley & Sons, NY.
Barrett, J. (1959) *Ph.D. thesis*, University of Manchester.
Bower, V.E. & Bates, R.W. (1955) *J. Res. Natn. Bur. Stand.*, **55**, 197
Chang, Y.Y. & Myerson, A.S. (1982) *Biotechnol. Bioeng.*, **24**, 889.
Ewart, D.K. (1990) *Ph.D. thesis*, University of London.
Kristensen, H.B., Salomon, A. & Kokholm, G. (1991) *Anal.Chem.*, **63**, 885A.
Lowry, O.H., Rosebrough, N.J., Farr, A.L. & Randall, R.J. (1957) *J.Biol.Chem.*, **193**, 265.
Muyzer, G.A., de Bruyn, A.C., Schmedding, D.J.M., Bos, P., Westbroek, P. & Kuenen, G.J. (1987) *Appl. Environ. Microbiol.*, **53**, 660.
Shrestha, G.N. (1988) *Australian Mining*, 48.
Skoog, D.A., West, D.M. & Holler, F.J. (1992) *Fundamentals of Analytical Chemistry.* 6th edn Saunders, NY.
Yeh, T.Y., Godshalk, J.R., Olson, G.J. & Kelly, R.M. (1987) *Biotechnol. Bioeng.*, **30**, 138.

Index

abundant metals, 29
acid mine drainage, 3, 10, 68
acid rain, 2
Acidanus, 5, 53
Acidanus brierleyi, 5, 111
acidophiles, 47
activity, 74
activity coefficient, 74
aeration, 131, 137
Africa, 25, 27, 28
agitated reactors, 134
agitated tank leaching, 9, 134, 151
Agnew Lake, 20
akaganeite, 157
aluminium, 29
amaranite, 158
America, 27, 28
ammonium ion, 5
analytical methods, 178
ancillary services, 170
antimony, 2, 4, 11, 26, 29, 30
archaebacteria, 53
arsenic, 6
arsenic(III) concentration, 124, 183
arsenic(III) oxide, 2
arsenic(V) acid, 2
arsenic(V) concentration, 124
arseniosiderite, 162
arsenobetaine, 167
arsenopyrite, 1, 3, 14
Asia, 25, 27, 28
atomic absorption
  spectroscopy, 179
atomic emission spectroscopy, 180
Australasia, 25
Australia, 13, 18, 20, 27, 28
autonite, 20
autotroph, 4

bacterial adaptation, 18

bacterial
  adhesion, 57, 108
  catalysis, 56, 109
  cells
    ecology, 47, 55
    metabolism, 42
    morphology, 39, 44
    physiology, 46, 54
    characteristics, 43
    DNA, 55
    growth, 62, 107
bacterial oxidation
  direct, 104
  indirect, 104
  methods, 12, 134, 151
  primary, 56, 110
  secondary, 116, 121
bacterial population, 113, 126
bacterial population in solution, 118
basic calcium-iron(III) arsenates, 162
basic iron(III) arsenate minerals, 160
basic iron(III) arsenate sulfate minerals, 161
basic iron(III) sulfates, 158
bimolecular collisions, 105
Bingham Canyon mine, 3
bioaccumulation, 38
biohydrometallurgy, 3, 4
biomass measurement, 184
Bisbee mine, 3
bismuth, 2, 30
Bolivia, 24, 26
bornite, 24
brannerite, 20
bravoite, 21
Brazil, 13, 24
bukovskyite, 161
butlerite, 158

cadmium, 30
calamine, 27

Index

calciocopiapite, 160
calcium arsenates, 162
calcium ion, 5
calomel electrode, 181
Canada, 13, 18, 19, 20, 23, 27
capital costs, 169
capsules, 42
carbon, 6
carbon dioxide, 4, 163
carnotite, 20
cassiterite, 23
catalysts, 3
cell metabolism, 42
cell wall, 40, 41, 45
chalcocite, 4, 24
chalcophiles, 30, 33
chalcopyrite, 1, 4, 16, 24
chemical dissolution of
    arsenopyrite, 101
    chalcopyrite, 100
    pyrite, 101
chemolithoautotroph, 4
chemolithotroph, 4
Chile, 23
China, 22, 26
chloride ion, 5
chlorine, 2
chromium, 29
class I process, 9, 14, 18
class II process, 9, 19, 22, 23, 26, 27, 29
class III process, 9, 19, 20, 26
classification of processes, 8
classifications of treatable elements, 29
cobalt, 2, 4, 19, 29, 30, 134
cobaltite, 4, 19
coffinite, 20
colorimetry, 183
construction materials, 170
copper, 2, 3, 4, 10, 11, 24, 29, 30, 134
covellite, 4, 24
cryophiles, 4, 5
Cuba, 22
culture growth, 133
cyanidation, 1, 18
cyanobacteria, 39
cytoplasm, 40, 41, 46
cytoplasmic membrane, 40, 41, 45

death phase, 108
Debye-Hückel equation, 75
digenite, 25
dimethylarsinic acid, 167
direct bacterial oxidation, 104
dissolution of gases, 106
djurleite, 25
doubling time, 62, 108
drippers, 145
dump leaching, 10, 152
dumps, 148

electrode potential, 125
electronegativity, 33
electronegativity coefficients, 33
electrowinning, 22, 155
Elliot Lake, 20
enargite, 24
enhancement of gold recovery, 6
environmental issues, 1
eucaryotes, 39
Europe, 20, 25, 27, 28
exometabolites, 60
exponential phase, 107
extreme thermophiles, 5

facultative chemolithotrophs, 5
Fermi level, 61
ferrisymplesite, 160
fibroferrite, 158
Finland, 19
flagella, 40
flow sheet
    bacterial oxidation, 140
    base metals, 141
    gold, 140

galena, 28
garnierite, 21
genus
    acidanus, 38
    leptospirillum, 38, 48
    metallosphaera, 38
    sulfobacillus, 38, 51
    sulfolobus, 5, 38
    sulfurococcus, 38
    thiobacillus, 38, 44
Gibbs energy, 73
glass electrode, 181
goethite, 158
gold, 1, 11, 13, 29, 30, 134
gold analysis, 184
Gram staining test, 39
growth phase, 107
gypsum, 161

hafnium, 31
heap leaching, 10, 152
heaps, 141
hematite, 158
heterogeneous processes, 105
heterotrophs, 4, 38
hohmannite, 158
hydrothermal activity, 31

igneous rocks, 30
impellers, 136
in situ leaching, 10, 20, 21, 26, 150, 152
INCO, 22
India, 28
indirect bacterial oxidation, 104
Indonesia, 24

Index

inductively coupled plasma
    mass spectrometry, 181
intermediates, 112
in-place leaching, 10
ion chromatography, 183
ionic strength, 75
iridium, 30
iron, 29
iron(III) arsenate minerals, 160
iron(III) concentration, 123, 183
iron(III) hydrolysis, 81
iron(III) oxohydroxide, 158
iron(II) concentration, 125
isolation of cultures, 68

jamesonite, 26
Japan, 27
jarosites, 87, 158
junction potentials, 181

kankite, 160
kinetic rate law, 105
kolfanite, 162

lag phase, 107
laterite, 21
lead, 2, 11, 27, 29, 30
lead(II) sulfate, 28
lepidocrocite, 158
Leptospirillum, 38
  ferrooxidans, 5, 48
  thermoferrooxidans, 49
liberation, 5
lime, 156
limestone, 156
limonite, 158
linnaeite, 19
liquid ion exchange, 22, 155
lithium hydroxide, 16
lithium monoxide, 21
lithophiles, 30, 31, 33
livingstonite, 26
lysis, 108

magnesium ion, 5
maintenance of cultures, 71
major equipment, 170
Malaysia, 24
manganese, 5, 29
marcasite, 15
marine muds, 31
maximum simplicity, 105
mean ionic activity coefficient, 75
mechanistic conclusions, 118
mechanistic principles, 105
mercury, 30
mesophile, 5, 38
metal liberation, 8
metallic iron, 26

metal-metal bonding, 21
methylarsonic acid, 167
Mexico, 18, 26
microbiology, 4, 39
millerite, 21
mixed cultures, 5, 65
mixed thermophilic culture, 53
moderate thermophiles, 5, 50
  non-classified, 52
molality, 74
molarity, 74
molybdenum, 2, 4, 22, 29, 30, 134
molybdic acid, 23

Namibia, 20
Nernst equation, 73
Neves Corvo mine, 24
niccolite, 21
nickel, 2, 4, 16, 21, 29, 30, 134
nickeliferous limonite, 21
nickeline, 21
Niger, 20
niobium, 31
nitrate ion, 5
nitric acid, 2
North America, 25
nutrients, 5, 69, 133

obligate aerobes, 5
obligate chemolithotrophs, 5
Occam's razor, 105
operating costs, 171
order of reaction, 105
osmium, 30
oxidation of
  arsenic(III), 98
  arsenopyrite, 97
  chalcopyrite, 96
  pyrite, 95
oxidizing power of
  iron(III), 93
  oxygen, 93

pachucas, 135
palladium, 30
parabutlerite, 158
particle size, 113, 129
pentlandite, 4, 16, 21, 33
periodic table, 32
Peru, 18
pH, 74, 123
pH measurement, 74, 181
pharacosiderite, 160
phases of bacterial growth, 107
Philippines, 13
phosphate ion, 5
pili, 40
pitchblende, 20
platinum, 30

Index

polarography, 183
pollution, 1
potassium, 29
potassium ion, 5
pressure oxidation, 2
primary mineral oxidation, 8
primary process, 56, 110
primary products, 116
procaryotes, 39
protein analysis, 185
protista, 39
psychrophiles, 4
pulp density, 113, 126, 129
pyrite, 1, 4, 15
pyrrhotite, 1, 16

rate determining step, 106
rate limiting, 106
recovery of gold, 154
recovery of solubilized metals, 155
reduction potentials, 61
refractory concentrates, 2
refractory gold, 1
refractory ores, 2
rhenium, 30
rhodium, 30
roasting, 1, 2, 174
role of iron(III), 112
ruthenium, 30

sarmienite, 161
scarce metals, 29
scorodite, 160
secondary mineral oxidation, 8
secondary oxidation of
   arsenic(III), 121
   iron(II), 121
secondary processes, 110, 116
sedimentary rocks, 31
semi-conduction, 14, 16, 110
shales, 31
siderophiles, 30, 33
silver, 1, 11, 18, 29, 30, 134
silver sulfide, 18
skutterudite, 19
slimes, 42
smaltite, 19
smithsonite, 27
sodium, 29
solute transport, 109
South Africa, 13, 20, 26, 27, 28
South America, 25
Spain, 27
speciation
   arsenic(III), 77
   arsenic(V), 78
   iron(III), 81
   iron(II), 80
   sulfur(VI), 76

sphalerite, 4, 27
stainless steel, 22
standard electrode potential, 73
stannite, 4, 24
stationary phase, 108
sterilization, 69
stibnite, 4, 26
storage of cultures, 71
sub-culturing procedure, 70
sulfate ion, 5
sulfide solubility, 35
Sulfobacillus, 5, 51
   thermosufidooxidans, 51
Sulfolobus, 5, 23, 53
   acidocaldarius, 63
sulfur dioxide, 2
sulfuric acid, 2
surface charge, 59
Sweden, 28
synergic interactions, 5

tank leaching, 151
tantalum, 31
temperature control, 132, 138
tetrahedrite, 26
Thailand, 24, 26
thermophiles, 5, 38
Thiobacillus, 38
   acidophilus, 48
   intermedius, 133
   ferrooxidans, 3, 38, 43, 44
   thiooxidans, 5, 43, 44
   thioparus, 47
tin, 2, 4, 11, 23, 29, 31, 134
titanium, 29
toxicity, 63
   arsenic compounds, 167
   molybdenum, 23
   silver, 18
   uranium, 21
troilite, 16
tungsten, 31
Turkey, 26

ullmannite, 21, 26
uraninite, 4, 20
uranium, 4, 11, 20, 29, 31, 134
uranyl cation, 20
USA, 13, 18, 20, 23, 28
USSR, 22, 26

vanadium, 29
vat leaching, 10, 152
vats, 149
violerite, 21
Wheal Jane mine, 3
wigglers, 145
wobblers, 145
wolfsbergite, 26

Index

world prices of metals, 13
world production of metals, 13

Zaire, 19
Zambia, 19

zinc, 2, 4, 5, 27, 29, 30
zinc blende, 27
zirconium, 31
zykaite, 161